Handbook of
Valves
Design, Selection and Uses

Handbook of Valves

Design, Selection and Uses

Walter Brown

Oxford Book Company

Jaipur ▪ New Delhi

Published By

Oxford Book Company

Regd. Off. "Bony Residency", Gate No. 2
Opp. Tilak Public School, Vishweswariya Nagar,
Gopalpura Road, Jaipur - 302018, India
Ph: 0141 - 2761280, Telefax: 2761381
email - oxfordbookjpr@gmail.com
website - oxfordbookcompany.com

Delhi Office: 102, 1st Floor, Satyam House,
4327/3, Ansari Road, Darya Ganj,
New Delhi - 110002, India
Ph: 011-45652440

Distributed by
KEYA INTERNATIONAL PRESS LLP
E-mail: keyainternationalpressllp@gmail.com

First published in India, 2024

ISBN : 9789355240699

Typeset by Shivangi Computers, Jaipur

Printed and bound in India by Replika Press Pvt. Ltd.

Contents

Preface

In the daily operation of chemical processing plant, overpressure can happen due to incidents like a blocked discharge, fire exposure, tube rupture, check valve failure, thermal expansion that can happen at process heat exchanger, and the failures can occur. This can lead to a major incident in plant if the pressure relief system is not in place or not functional. Is very important to properly select, size, locate and maintain the pressure relief systems to prevent or minimize the losses from major incident like fire or other issues. Detail of selection and sizing of pressure relief valve is illustrated in the following sections. Pressure relief system is used to protect piping and equipment against excessive overpressure for equipment and personnel safety. Pressure relief systems consist of a pressure relief device, flare piping system, flare separation drum and flare system.

A pressure relief device is designed to open and relieve excess pressure; it is re-closed after normal conditions have been restored to prevent the further flow of fluid (except for a rupture disk). Overpressures that occur in chemical plants and refineries have to be reviewed and studied, it is important in preliminary steps of pressure relief system design. It helps the designer to understand the cause of overpressure and to minimize the effect. Overpressure is the result of an unbalance or disruption of the normal flows of material and energy that causes the material or energy, or both, to build up in some part of the system. As mentioned earlier, blocked discharge, fire exposure, tube rupture, check valve failure, thermal expansion happen at process line heat exchanger, and utility failure can cause over pressure in process equipment. Blocked discharge can be defined as any vessel, pump, compressor, fired heater, or other equipment item which closure of block valve at outlet either by mechanical failure or human error. This will expose the vessel to a pressure that exceeds the maximum allowable working pressure, and a pressure relief device is required unless administrative procedures to control valve closure such as car seals or locks are in place.

A vessel which normally contains no liquid, since failure of the shell from overheating would probably occur even if a pressure relief valve were provided. Vessel (drums or towers) with 2 ft or less in diameter, constructed of pipe, pipe fittings or equivalent, do not require pressure relief valves for protection against fire, unless these are stamped as coded vessels. Heat exchangers do not need a separate pressure relief valve for protection against fire exposure since they are usually protected by pressure relief valves in interconnected equipment or have an open escape path to atmosphere via a cooling tower or tank. Vessels filled with both a liquid and a solid (such as molecular sieves or catalysts) not require pressure relief valve for protection against fire exposure. In this case, the behaviour of the vessel contents normally precludes the cooling effect of liquid boiling. Hence rupture discs, fireproofing and de ressuring should be considered as alternatives to protection by pressure relief valves.

This book meets the need of engineers who have responsibilities for specifying, installing, inspecting or maintaining safety valves and flow control systems.

— Walter Brown

Chapter 1

Introduction

A valve is a device that regulates, directs or controls the flow of a fluid (gases, liquids, fluidized solids, or slurries) by opening, closing, or partially obstructing various passageways. Valves are technically pipe fittings, but are usually discussed as a separate category. In an open valve, fluid flows in a direction from higher pressure to lower pressure.

The simplest, and very ancient, valve is simply a freely hinged flap which drops to obstruct fluid (gas or liquid) flow in one direction, but is pushed open by flow in the opposite direction.

Valves are used in a variety of contexts, including industrial, military, commercial, residential, and transport. The industries in which the majority of valves are used are oil and gas, power generation, mining, water reticulation, sewage and chemical manufacturing.

In daily life, most noticeable are plumbing valves, such as taps for tap water. Other familiar examples include gas control valves on cookers, small valves fitted to washing machines and dishwashers, safety devices fitted to hot water systems, and valves in car engines. In nature, veins acting as valves are controlling the blood circulation; heart valves control the flow of blood in the chambers of the heart and maintain the correct pumping action.

Valves play a vital role in industrial applications ranging from transportation of drinking water to control of ignition in a rocket engine.

Valves may be operated manually, either by a handle, lever or pedal. Valves may also be automatic, driven by changes in pressure, temperature, or flow. These changes may act upon a diaphragm or a piston which in turn activates the valve, examples of this type of valve found commonly are safety valves fitted to hot water systems or boilers.

More complex control systems using valves requiring automatic control based on an external input (i.e., regulating flow through a pipe to a changing set point) require an actuator. An actuator will stroke the valve depending on its input and set-up, allowing the valve to be positioned accurately, and allowing control over a variety of requirements.

Applications

Valves vary widely in form and application. Sizes typically range from 0.1 mm to 60 cm. Special valves can have a diametre exceeding 5 metres. Valve cost ranges from simple inexpensive disposable valves to specialised valves cost thousands of US dollars per inch of diametre. Disposable valves may be found inside common household items including mini-pump dispensers and aerosol cans.

Types of Valves: Valves are quite diverse and may be classified into a number of basic types. Valves may also be classified by how they are actuated:

Ball Valve

A ball valve is a valve with a spherical disc, the part of the valve which controls the flow through it. The sphere has a hole, or port, through the middle so that when the port is in line with both ends of the valve, flow will occur. When the valve is closed, the hole is perpendicular to the ends of the valve, and flow is blocked. The handle or lever will be inline with the port position letting you "see" the valve's position. The ball valve, along with the butterfly valve and plug valve, are part of the family of *quarter turn valves*.

Ball valves are durable and usually work to achieve perfect shutoff even after years of disuse. They are therefore an excellent choice for shutoff applications (and are often preferred to globe valves and gate valves for this purpose). They do not offer the fine control that may be necessary in throttling applications but are sometimes used for this purpose. Ball valves are used extensively in industrial applications because they are very versatile, supporting pressures up to 1000 bars and temperatures up to 482°F (250°C). Sizes typically range from 0.2 to 11.81 inches (0.5 cm to 30 cm). They are easy to repair and operate. The body of ball valves may be made of metal, plastic or metal with a ceramic centre. The ball is often chrome plated to make it more durable. A ball-check valve is a type of check valve with a ball without a hole for a disc, and is not covered in this article. The genericized trademark *ball-o-fix* is occasionally used after the original UK market leader.

The Ballofix was invented by the Danish company Broen, and are still in production.

Types of Ball Valve: There are five general body styles of ball valves: single body, three piece body, split body, top entry, and welded. The difference is based on how the pieces of the valve—especially the casing that contains the ball itself—are manufactured and assembled. The valve operation is the same in each case.

Figure: Duplex ball valve

In addition, there are different styles related to the bore of the ball mechanism itself:

- A *full port* or more commonly known *full bore* ball valve has an over-sized ball so that the hole in the ball is the same size as the pipeline resulting in lower friction loss. Flow is unrestricted but the valve is larger and more expensive so this is only used where free flow is required, for example in pipelines which require pigging.
- In *reduced port* or more commonly known *reduced bore* ball valves, flow through the valve is one pipe size smaller than the valve's pipe size resulting in flow area being smaller than pipe. As the flow discharge remains constant and is equal to area of flow (A) times velocity (V), $A_1V_1 = A_2V_2$ the velocity increases with reduced area of flow.
- A *V port* ball valve has either a 'v' shaped ball or a 'v' shaped seat. This allows the orifice to be opened and closed in a more controlled manner with a closer to linear flow characteristic. When the valve is in the closed position and opening is commenced the small end of the 'v' is opened first allowing stable flow control during this stage.
- A trunnion ball valve has additional mechanical anchoring of the ball at the top and the bottom, suitable for larger and higher pressure valves (say, above 10 cm and 40 bars).

- Cavity filler Ball Valve. Many industries encounter problem with residues in the ball valve. Where the fluid is meant for human consumption, residues may also be health hazard, and when where the fluid changes from time to time contamination of one fluid with another may occur. Residues arise because in the half open position of the ball valve a gap is created between the ball bore and the body in which fluid can be trapped. To avoid the fluid getting into this cavity, the cavity has to be plugged, which can be done by extending the seats in such a manner that it is always in contact with the ball. This type of ball valve is known as Cavity Filler Ball Valve.

Manually operated ball valves can be closed quickly and thus there is a danger of water hammer. Some ball valves are equipped with an actuator that may be pneumatically or motor operated. These valves can be used either for on/off or flow control. A pneumatic flow control valve is also equipped with a positioner which transforms the control signal into actuator position and valve opening accordingly.

Three-way and four-way ball valves

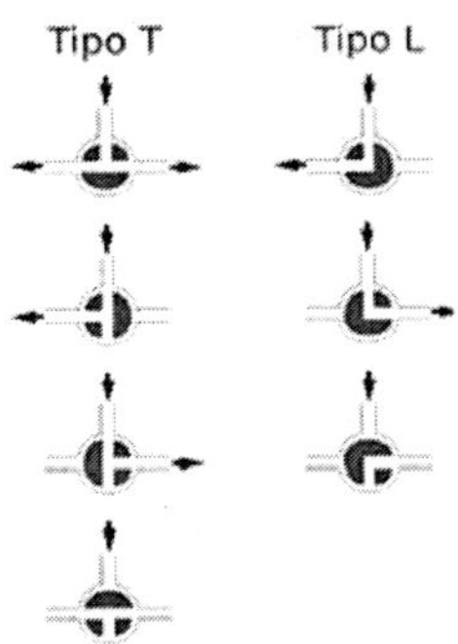

Figure: Schematic 3 way ball valve: L-shaped ball right, T-shaped left

Three-way ball valves have an L- or T-shaped hole through the middle. The different combinations of flow are shown in the figure. It is easy to see that a T valve can connect any pair of ports, or all three, together, but the 45 degree position which might disconnect all three leaves no margin for error. The L valve can connect the centre port to either side port, or disconnect all three, but it cannot connect the side ports together.

Multi-port ball valves with 4 ways, or more, are also commercially available, the inlet way often being orthogonal to the plane of the outlets. For special applications, such as driving air-powered motors from forward to reverse, the operation is performed by rotating a single lever four-way valve. The 4-way ball valve has two L-shaped ports in the ball that do not interconnect, sometimes referred to as an "×" port.

Ball valves in sizes up to 2 inch generally come in single piece, two or three piece designs. One piece ball valves are almost always reduced bore, are relatively inexpensive and generally are throw-away. Two piece ball valves are generally slightly reduced (or standard) bore, they can be either throwaway or repairable. The 3 piece design allows for the centre part of the valve containing the ball, stem & seats to be easily removed from the pipeline. This facilitates efficient cleaning of deposited sediments, replacement of seats and gland packings, polishing out of small scratches on the ball, all this without removing the pipes from the valve body. The design concept of a three piece valve is for it to be repairable.

Butterfly Valve

A butterfly valve is a valve which can be used for isolating or regulating flow. The closing mechanism takes the form of a disk. Operation is similar to that of a ball valve, which allows for quick shut off. Butterfly valves are generally favoured because they are lower in cost to other valve designs as well as being lighter in weight, meaning less support is required. The disc is positioned in the centre of the pipe, passing through the disc is a rod connected to an actuator on the outside of the valve. Rotating the actuator turns the disc either parallel or perpendicular to the flow. Unlike a ball valve, the disc is always present within the flow, therefore a pressure drop is always induced in the flow, regardless of valve position.

A butterfly valve is from a family of valves called quarter-turn valves. The "butterfly" is a metal disc mounted on a rod. When the valve is closed, the disc is turned so that it completely blocks off the passageway. When the valve is fully open, the disc is rotated a quarter turn so that it allows an almost unrestricted passage of the fluid. The valve may also be opened incrementally to throttle flow.

There are different kinds of butterfly valves, each adapted for different pressures and different usage. The resilient butterfly valve, which uses the flexibility of rubber, has the lowest pressure rating. The high performance butterfly valve, used in slightly higher-pressure systems, features a slight offset in the way the disc is positioned, which increases the valve's sealing ability and decreases its tendency to wear. The valve best suited for high-pressure systems is the triple offset butterfly valve, which makes use of a metal seat, and is therefore able to withstand a greater amount of pressure.

Structure

Butterfly valves are valves with a circular body and a rotary motion disk closure member which is pivotally supported by its stem. A butterfly valve can appear in various styles, including eccentric and high-performance valves. These are normally a type of valve that uses a flat plate to control the flow of water. As well as this, butterfly valves are used on firefighting apparatus and typically are used on larger lines, such as front and rear suction ports and tank to pump lines.

Types:

1. *Concentric butterfly valves* - This type of valves has a resilient rubber seat with a metal disc.
2. *Doubly-eccentric butterfly valves* (*high-performance butterfly valves* or *double-offset butterfly valves*) - Different type of materials is used for seat and disc.
3. *Triply-eccentric butterfly valves* (*triple-offset butterfly valves*) - The seats are either laminated or solid metal seat design.

Triply-Eccentric (Offset) Design

Triple eccentric (offset) Design prevents galling and scratches between the metal seat and the metal disc due to its unique design. The only time where the seal comes into contact with the seat is at the point of complete closure. Triple offset valves are generally used in applications which require bi-directional tight shut-off in oil and gas, LNG/NPG terminal and tanks, chemical factories, and shipbuilding. Widely use for dirty/heavy oil to prevent extrusion.

Wafer-Style Butterfly Valves

The wafer style butterfly valve is designed to maintain a seal against bi-directional pressure differential to prevent backflow in systems designed for unidirectional flow. It accomplishes this with a tightly fitting seal; i.e., gasket, o-ring, precision machined, and a flat valve face on the upstream and downstream sides of the valve.

Lug-Style Butterfly Valve

Lug-style valves have threaded inserts at both sides of the valve body. This allows them to be installed into a system using two sets of bolts and no nuts. The valve is installed between two flanges using a separate set of bolts for each flange. This setup permits either side of the piping system to be disconnected without disturbing the other side. A lug-style butterfly valve used in dead end service generally has a reduced pressure rating. For example a lug-style butterfly valve mounted between two flanges has a 150 psi pressure rating. The same valve mounted with one flange, in dead end service, has a 75 psi rating.

Check Valve

A check valve, clack valve, non-return valve or one-way valve is a mechanical device, a valve, which normally allows fluid (liquid or gas) to flow through it in only one direction.

Check valves are two-port valves, meaning they have two openings in the body, one for fluid to enter and the other for fluid to leave. There are various types of check valves used in a wide variety of applications. Check valves are often part of common household items. Although they are available in a wide range of sizes and costs, check valves generally are very small, simple, and/or inexpensive. Check valves work automatically and most are not controlled by a person or any external control; accordingly, most do not have any valve handle or stem. The bodies (external shells) of most check valves are made of plastic or metal.

An important concept in check valves is the cracking pressure which is the minimum upstream pressure at which the valve will operate. Typically the check valve is designed for and can therefore be specified for a specific cracking pressure.

Heart valves are essentially inlet and outlet check valves for the heart ventricles, since the ventricles act as pumps.

Types of Check Valves

Backwater valve (for sanitary drainage system) protects lower located rooms against flooding caused by return flow of sewage waters. Such risk occurs most often in sanitary drainage systems connected to combined sewerage systems and in rainwater drainage systems. It may be caused by intense rainfall, thaw or flood. Backwater valve prevents rats and other rodents entering the sanitary and rainwater drainage systems and, consequently, the building interiors. It protects also against unpleasant smells in case of longer breaks in system use.

A ball check valve is a check valve in which the closing member, the movable part to block the flow, is a spherical ball. In some ball check valves, the ball is spring-loaded to help keep it shut. For those designs without a spring, reverse flow is required to move the ball toward the seat and create a seal. The interior surface of the main seats of ball check valves are more or less conically-tapered to guide the ball into the seat and form a positive seal when stopping reverse flow.

Ball check valves are often very small, simple, and cheap. They are commonly used in liquid or gel minipump dispenser spigots, spray devices, some rubber bulbs for pumping air, etc., manual air pumps and some other pumps, and refillable dispensing syringes. Although the balls are most often made of metal, they can be made of other materials, or in some specialised cases out of artificial ruby. High pressure HPLC pumps and similar applications commonly use small inlet and outlet ball check valves with both balls and seats made of artificial ruby, for both hardness and chemical resistance. After prolonged use, such check valves can eventually wear out or the seat can develop a crack, requiring replacement. Therefore, such valves are made to be replaceable, sometimes placed in a small plastic body tightly-fitted inside a metal fitting which can withstand high pressure and which is screwed into the pump head.

There are similar check valves where the disc is not a ball, but some other shape, such as a poppet energized by a spring. Ball check valves should not be confused with ball valves, which is a different type of valve in which a ball acts as a controllable rotor to stop or direct flow. A diaphragm check valve uses a flexing rubber diaphragm positioned to create a normally-closed valve. Pressure on the upstream side must be greater than the pressure on the downstream side by a certain amount, known as the pressure differential, for the check valve to open allowing flow. Once positive pressure stops, the diaphragm automatically flexes back to its original closed position.

A swing check valve or tilting disc check valve is check valve in which the disc, the movable part to block the flow, swings on a hinge or trunnion, either onto the seat to block reverse flow or off the seat to allow forward flow. The seat opening cross-section may be perpendicular to the centreline between the two ports or at an angle. Although swing check valves can come in various sizes, large check valves are often swing check valves.

The flapper valve in a flush-toilet mechanism is an example of this type of valve. Tank pressure holding it closed is overcome by manual lift of the flapper. It then remains open until the tank drains and the flapper falls due to gravity. Another variation of this mechanism is the clapper valve, used in applications such firefighting and fire life safety systems. A hinged gate only remains open in the inflowing direction. The clapper valve often also has a spring that keeps the gate shut when there is no forward pressure.

A stop-check valve is a check valve with override control to stop flow regardless of flow direction or pressure. In addition to closing in response to backflow or insufficient forward pressure (normal check-valve behaviour), it can also be deliberately shut by an external mechanism, thereby preventing any flow regardless of forward pressure.

A lift-check valve is a check valve in which the disc, sometimes called a lift, can be lifted up off its seat by higher pressure of inlet or upstream fluid to allow flow to the outlet or downstream side. A guide keeps motion of the disc on a vertical line, so the valve can later reseat properly. When the pressure is no longer higher, gravity or higher downstream pressure will cause the disc to lower onto its seat, shutting the valve to stop reverse flow.

A duckbill valve is a check valve in which flow proceeds through a soft tube that protrudes into the downstream side. Back-pressure collapses this tube, cutting off flow.

Multiple check valves can be connected in series. For example, a double check valve is often used as a backflow prevention device to keep potentially contaminated water from siphoning back into municipal water supply lines. There are also double ball check valves in which there are two ball/seat combinations sequentially in the same body to ensure positive leak-tight shutoff when blocking reverse flow; and piston check valves, wafer check valves, and ball-and-cone check valves.

Applications

Pumps

The check valves on this steam locomotive are located under the small dome between the chimney and the main dome

Check valves are often used with some types of pumps. Piston-driven and diaphragm pumps such as metring pumps and pumps for chromatography commonly use inlet and outlet ball check valves. These valves often look like small cylinders attached to the pump head on the inlet and outlet lines. Many similar pump-like mechanisms for moving volumes of fluids around use check valves such as ball check valves. The feed pumps or injectors which supply water to steam boilers are fitted with check valves to prevent back-flow.

Industrial Processes

Check valves are used in many fluid systems such as those in chemical and power plants, and in many other industrial processes.

Check valves are also often used when multiple gases are mixed into one gas stream. A check valve is installed on each of the individual gas streams to prevent mixing of the gases in the original source. For example, if a fuel and an oxidizer are to be mixed, then check valves will normally be used on both the fuel and oxidizer sources to ensure that the original gas cylinders remain pure and therefore nonflammable.

Domestic Use

Some types of irrigation sprinklers and drip irrigation emitters have small check valves built into them to keep the lines from draining when the system is shut off.

Also used with most home made snowmakers.

Check valves used in domestic heating systems to prevent vertical convection, especially in combination with solar thermal installations, also are called gravity brake.

Rainwater harvesting systems that are plumbed into the main water supply of a utility provider may be required to have one or more check values fitted to prevent contamination of the primary supply by rainwater.

History

Frank P. Cotter developed a "simple self sealing check valve, adapted to be connected in the pipe connections without requiring special fittings and which may be readily opened for inspection or repair" 1907 (U.S. patent #865,631).

Nikola Tesla invented a deceptively simple one-way valve for fluids in 1916, called a Tesla valve. It was patented in 1920 (U.S. patent 1,329,559).

Duckbill Valve

A duckbill valve is a valve, manufactured from rubber or synthetic elastomer, and shaped like the beak of a duck. It is commonly used in medical applications to prevent contamination due to backflow.

One end of the valve is stretched over the outlet of a supply line, conforming itself to the shape of the line, usually round. The other end, the duckbill, retains its natural flattened shape. When a fluid is pumped through the supply line and therefore the duckbill, the flattened end opens to permit the pressurized fluid to pass. When pressure is removed, however, the duckbill end returns to its flattened shape, preventing backflow.

The duckbill valve is similar in function to the mitral valve in the heart.

Mitral Valve

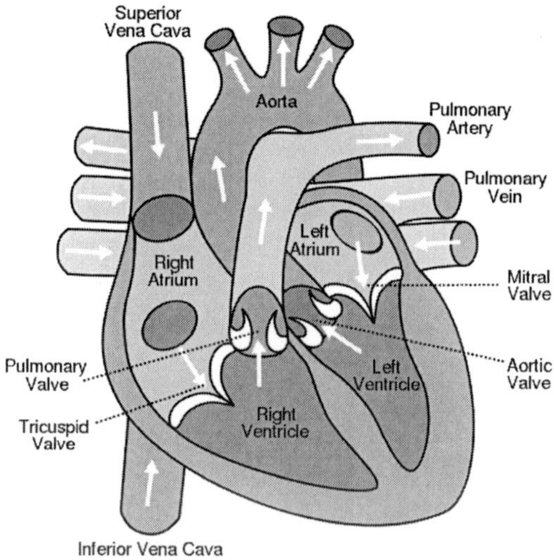

Figure: Anterior (frontal) view of the opened heart. White arrows indicate normal blood flow. (Mitral valve labelled at centre right.)

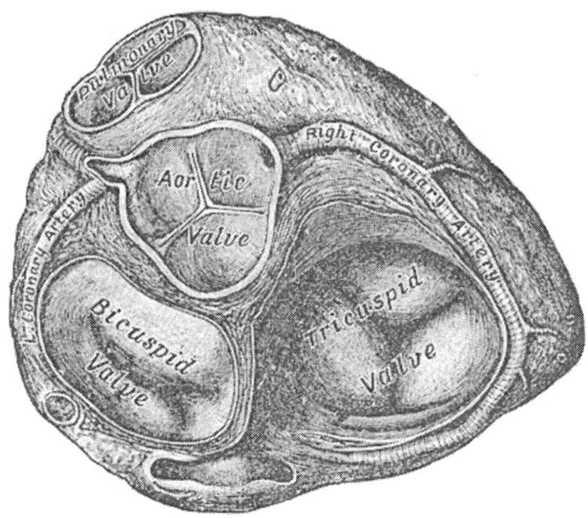

Figure: Base of ventricles exposed by removal of the atria. (Bicuspid (mitral) valve visible at bottom left.)

The mitral valve (also known as the bicuspid valve or left atrioventricular valve) is a dual-flap (bi- from the Latin, meaning *double*, and mitral- from the Latin, meaning *shaped like a mitre*) valve in the heart that lies between the left atrium (LA) and the left ventricle (LV). The mitral valve (not to be confused with the congenital bicuspid aortic valve) and the tricuspid valve are known collectively as the atrioventricular valves because they lie between the atria and the ventricles of the heart and control the flow of blood.

During diastole, a normally-functioning mitral valve opens as a result of increased pressure from the left atrium as it fills with blood (preloading). As atrial pressure increases above that of the left ventricle, the mitral valve opens. Opening facilitates the passive flow of blood into the left ventricle. Diastole ends with atrial contraction, which ejects the final 20% of blood that is transferred from the left atrium to the left ventricle. This amount of blood is known as end diastolic volume (EDV), and the mitral valve closes at the end of atrial contraction to prevent a reversal of blood flow.

Anatomy

Operative view of The Mitral valve with a chordal rupture "fail" of the anterior leaflet

The mitral valve is typically 4–6 cm^2 in area. It has two cusps, or leaflets, (the *anteromedial leaflet* and the *posterolateral leaflet*) that guard the opening. The opening is surrounded by a fibrous ring known as the *mitral valve annulus*. (The orientation of the two leaflets resemble a bishop's mitre, whence the valve receives its name.) The anterior cusp protects approximately two-thirds of the valve (imagine a crescent moon within the circle, where the crescent represents the posterior cusp). Note that although the anterior leaflet takes up a larger part of the ring and rises higher, the posterior leaflet has a larger surface area. These valve leaflets are prevented from prolapsing into the left atrium by the action of tendons attached to the posterior surface of the valve, chordae tendineae.

The inelastic chordae tendineae are attached at one end to the papillary muscles and the other to the valve cusps. Papillary muscles are fingerlike projections from the wall of the left ventricle. Chordae tendineae from each muscle are attached to both leaflets of the mitral valve. Thus, when the left ventricle contracts, the intraventricular pressure forces the valve to close, while the tendons keep the leaflets coapting together and prevent the valve from opening in the wrong direction (thus preventing blood to flow back to the left atrium). Each chord has a different thickness. The thinnest ones are attached to the free leaflet margin, whereas thickest ones (strut chords) are attached quite away from the free margin. This disposition has important effects on systolic stress distribution physiology (video).

Normal Physiology

During left ventricular diastole, after the pressure drops in the left ventricle due to relaxation of the ventricular myocardium, the mitral valve opens, and blood travels from the left atrium to the left ventricle. About 70-80% of the blood that travels across the mitral valve occurs during the early filling phase of the left ventricle. This early filling phase is due to active relaxation of the ventricular myocardium, causing a pressure gradient that allows a rapid flow of blood from the left atrium, across the mitral valve. This early filling across the mitral valve is seen on doppler echocardiography of the mitral valve as the E wave.

After the E wave, there is a period of slow filling of the ventricle.

Left atrial contraction (left atrial systole) (during left ventricular diastole) causes added blood to flow across the mitral valve immediately before left ventricular systole. This late flow across the open mitral valve is seen on doppler echocardiography of the mitral valve as the A wave. The late filling of the LV contributes about 20% to the volume in the left ventricle prior to ventricular systole, and is known as the atrial kick. The mitral annulus changes in shape and size during the cardiac cycle. It is smaller at the end of atrial systole due to the contraction of the left atrium around it, like a sphincter. This reduction in annulus size at the end of atrial systole may be important for the proper coapting of the leaflets of the mitral valve when the left ventricle contracts and pumps blood. Leaking valves can be corrected by mitral valve annuloplasty, a common surgical procedure that aims at restoring proper leaflet coaptation.

Surface Anatomy

The closing of the mitral valve and the tricuspid valve constitutes the first heart sound (S1). It is not actually the valve closure which produces a sound but rather the sudden cessation of blood flow caused by the closure of the mitral and tricuspid valves. The mitral valve opening is normally not heard except in Mitral Stenosis (narrowing of the valve) as the Opening Snap. Flow of blood into the heart during rapid filling is not normally heard except in certain pathological states where it constitutes the third heart sound (S3).

Choke Valve

A Choke valve is a type of valve designed to create choked flow in a fluid. Over a wide range of valve settings the flow through the valve can be understood by ignoring the viscosity of the fluid passing through the valve; the rate of flow is determined only by the ambient pressure on the upstream side of the valve.

In automotive contexts, a choke valve modifies the air pressure in the intake manifold of an internal combustion engine, thereby altering the ratio of fuel and air quantity entering the engine. Choke valves are generally used in naturally aspirated engines with carburetors, to supply a richer fuel mixture when starting the engine. Most choke valves in engines are butterfly valves mounted in the manifold above the carburetor jet, to produce a higher partial vacuum and thereby draw more fuel into the intake stream.

In heavy industrial or fluid engineering contexts, a choke valve is a particular design of valve that raises and lowers a solid cylinder (called a "plug" or "stem") which is placed around or inside another cylinder that has holes or slots. The design of a choke valve means fluids flowing through the cage are coming from all sides and that the streams of flow (through the holes or slots) collide with each other at the centre of the cage cylinder, thereby dissipating the energy of the fluid through "flow impingement". The main advantage of choke valves is that they can be designed to be totally linear in their flow rate.

Automotive

A choke valve is sometimes installed in the carburetor of internal combustion engines. Its purpose is to restrict the flow of air, thereby enriching the fuel-air mixture while starting the engine. Depending on engine design and application, the valve can be activated manually by the operator of the engine (via a lever or pull handle) or automatically by a temperature-sensitive mechanism called an autochoke.

Choke valves are important for normally aspirated gasoline engines because small droplets of gasoline do not evaporate well within a cold engine. By restricting the flow of air into the throat of the carburetor, the choke valve reduces the pressure inside the throat, which causes a proportionally greater amount of fuel to be pushed from the main jet into the combustion chamber during cold-running operation. Once the engine is warm (from combustion), opening the choke valve restores the carburetor to normal operation, supplying fuel and air in the correct stoichiometric ratio for clean, efficient combustion.

Note that the term "choke" is applied to the carburetor's enrichment device even when it works by a totally different method. Commonly SU carburetors have "chokes" that work by lowering the fuel jet to a narrower part of the needle. Some others work by introducing an additional fuel route to the constant depression chamber.

Chokes were nearly universal in automobiles until fuel injection began to supplant carburetors. Choke valves are still common in other internal-combustion applications, including most small portable engines, motorcycles, small propeller-driven airplanes, riding lawn mowers, and normally aspirated marine engines.

Industrial

Heavy duty industrial choke valves control the flow to a certain Flow Coefficient (Cv) determined by how far the valve is opened. They are regularly used in the oil industry and for highly erosive and corrosive purposes, they are often made of tungsten carbide or inconel.

Diaphragm Valve

Diaphragm valves (or membrane valves) consists of a valve body with two or more ports, a diaphragm, and a "saddle" or seat upon which the diaphragm closes the valve. The valve is constructed from either plastic or steel.

Originally, the diaphragm valve was developed for use in non-hygienic applications. Later on the design was adapted for use in the bio-pharmaceutical industry by using compliant materials that can withstand sanitizing and sterilizing methods.

There are two main categories of diaphragm valves: one type seals over a "weir" (saddle) and the other (sometimes called a "straight-way" valve) seals over a seat. The main difference is that a saddle-type valve has its two ports in line with each other on the opposite sides of the valve, whereas the seat-type has the in/out ports located at a 90 degree angle from one another. The saddle type is the most common in process applications and the seat-type is more commonly used as a tank bottom valve but exists also as a process valve. While diaphragm valves usually come in two-port forms (2/2-way diaphragm valve), they can also come with three ports (3/2-way diaphragm valves also called t-valves) and more (so called block-valves). When more than three ports are included, they generally require more than one diaphragm seat; however, special dual actuators can handle more ports with one membrane. Diaphragm valves can be manual or automated. Their application is generally as shut-off valves in process systems within the food and beverage, pharmaceutical and biotech industries. The older generation of these valves is not suited for regulating and controlling process flows, however newer developments in this area have successfully tackled this problem.

In addition to the well known, two way shut off diaphragm valve, other types include: three way zero deadleg valve, sterile access port, block and bleed, valbow and tank bottom valve just to name a few.

Actuators

Diaphragm valves can be controlled by various types of actuators e.g. manual, pneumatic, hydraulic, electric etc. The most common diaphragm valves use pneumatic actuators; in this type of valve, air pressure is applied through a Schrader valve which raises the diaphragm and opens the valve. This type of valve is extremely quick and as such is one of the more common valves used in operations where valve speed is a necessity. Hydraulic diaphragm valves also exist for higher pressure and lower speed operations. Some diaphragm valves are also controlled manually.

Body Materials

- Brass
- Steel type:
  - Cast Iron
  - Ductile Iron
  - Carbon Steel
  - Stainless Steel
- Plastic type:
  - ABS (Acrylonitrile butadiene styrene)
  - PVC-U (Polyvinyl chloride, unplasticized) also known as PVCu or uPVC
  - PVC-C (Polyvinyl chloride, post chlorinated) also known as PVCc or cPVC
  - PP (Polypropylene)
  - PE (Polyethylene) also known as LDPE, MDPE and HDPE
  - PVDF (Polyvinylidene fluoride)

Body Lining Materials

Depending on temperature, pressure and chemical resistance, one of the following is used:

- Unlined type
- Rubber lined type:
  - o NR/Hard Rubber/Ebonite,
  - o BR/Soft rubber
  - o EPDM
- Fluorine plastic lined type
  - o FEP/F46
  - o PFA
  - o PO

Diaphragm Materials

- Unlined or Rubber Lined Type:
  - o NR/Natural Rubber
  - o NBR/Nitrile/Buna-N
  - o EPDM
  - o FKM/Viton
  - o SI/Silicone rubber
- Fluorine Plastic Type:
  - o FEP/F46,with EPDM back
  - o PTFE/F4, with EPDM back
  - o PFA, with EPDM back

Diaphragm Valve Manufacturers

- Hills-McCanna - First in USA since 1931
- SED Flowcontrol, Germany
- Alfa Laval
- Bürkert
- Georg Fischer (Swiss company)
- Gemü
- Festo
- ITT Corporation
- Novaseptic
- Saunders
- Galassi & Ortolani SpA
- Kim Diaphragm Valves Australia.

Gate Valve

The gate valve, also known as a sluice valve, is a valve that opens by lifting a round or rectangular gate/wedge out of the path of the fluid. The distinct feature of a gate valve is the sealing surfaces between the gate and seats are planar, so gate valves are often used when a straight-line flow of fluid and minimum restric-tion is desired. The gate faces can form a wedge shape or they can be parallel. Gate valves are primarily used to permit or prevent the flow of liquids, but typical gate valves shouldn't be used for regulating flow, unless they are specifically designed for that purpose. Because of their ability to cut through liquids, gate valves are often used in the petroleum industry. On opening the gate valve, the flow path is enlarged in a highly nonlinear manner with respect to percent of opening. This means that flow rate does not change evenly with stem travel. Also, a partially open gate disk tends to vibrate from the fluid flow. Most of the flow change occurs near shutoff with a relatively high fluid velocity causing disk and seat wear and eventual leakage if used to regulate flow. Typical gate valves are designed to be fully opened or closed. When fully open, the typical gate valve has no obstruction in the flow path, resulting in very low friction loss.

Gate valves are characterised as having either a rising or a nonrising stem. Rising stems provide a visual indication of valve position because the stem is attached to the gate such that the gate and stem rise and lower together as the valve is operated. Nonrising stem valves may have a pointer threaded onto the upper end of the stem to indicate valve position, since the gate travels up or down the stem on the threads without raising or lowering the stem. Nonrising stems are used underground or where vertical space is limited.

Bonnets provide leakproof closure for the valve body. Gate valves may have a screw-in, union, or bolted bonnet. Screw-in bonnet is the simplest, offering a durable, pressure-tight seal. Union bonnet is suitable for applications requiring frequent inspection and cleaning. It also gives the body added strength. Bolted bonnet is used for larger valves and higher pressure applications. Another type of bonnet construction in a gate valve is pressure seal bonnet. This construction is adopted for valves for high pressure service, typically in excess of 15 MPa (2250 psi). The unique feature about the pressure seal bonnet is that the body - bonnet joints seals improves as the internal pressure in the valve increases, compared to other constructions where the increase in internal pressure tends to create leaks in the body-bonnet joint.

Gate valves may have flanged ends which are drilled according to pipeline compatible flange dimensional standards. Gate valves are typically constructed from cast iron, ductile iron, cast carbon steel, gun metal, stainless steel, alloy steels, and forged steels.

Globe Valve

A globe valve is a type of valve used for regulating flow in a pipeline, consisting of a movable disk-type element and a stationary ring seat in a generally spherical body.

Globe valves are named for their spherical body shape with the two halves of the *body* being separated by an internal baffle. This has an opening that forms a *seat* onto which a movable *plug* can be screwed in to close (or shut) the valve. The plug is also called a *disc* or *disk*. In globe valves, the plug is connected to a *stem* which is operated by screw action in manual valves. Typically, automated valves use sliding stems. Automated globe valves have a smooth stem rather than threaded and are opened and closed by an actuator assembly. When a globe valve is manually operated, the stem is turned by a handwheel. Although globe valves in the past had the spherical bodies which gave them their name, many modern globe valves do not have much of a spherical shape. However, the term *globe valve* is still often used for valves that have such an internal mechanism. In plumbing, valves with such a mechanism are also often called stop valves since they don't have the global appearance, but the term *stop valve* may refer to valves which are used to stop flow even when they have other mechanisms or designs. Globe valves are used for applications requiring throttling and frequent operation. For example, globe valves or valves with a similar mechanism may be used as sampling valves, which are normally shut except when liquid samples are being taken. Since the baffle restricts flow, they're not recommended where full, unobstructed flow is required.

Parts of a Typical Globe Valve

Body: The body is the main pressure containing structure of the valve and the most easily identified as it forms the mass of the valve. It contains all of the valve's internal parts that will come in contact with the substance being controlled by the valve. The bonnet is connected to the body and provides the containment of the fluid, gas, or slurry that is being controlled. Globe valves are typically two-port valves, although three port valves are also produced mostly in straight-flow configuration. *Ports* are openings in the body for fluid flowing in or out. The two ports may be oriented straight across from each other or anywhere on the body, or oriented at an angle (such as a 90°). Globe valves with ports at such an angle are called angle globe valves. Angle type globe valves are mainly used for corrosive or high viscous fluids which solidify at room temperature. This is because straight valves are designed so that the outlet pipe is in line with the inlet pipe and the fluid has a good chance of staying there in the case of horizontal piping.

In the case of angle valves, the outlet pipe is directed towards the bottom. This allows the fluid to drain off. In turn, this prevents clogging and/or corrosion of the valve components over a period of time. A globe valve can also have a body in the shape of a "Y". This will allow the construction of the valve to be straight at the bottom as opposed to the conventional pot type construction (to arrange bottom seat) in case of other valves. This will again allow the fluid to pass through without difficulty and minimises fluid clogging/ corrosion in the long term.

Bonnet

The bonnet provides a leakproof closure for the valve body. The threaded section of the stem goes through a hole with matching threads in the bonnet. Globe valves may have a screw-in, union, or bolted[5] bonnet. Screw-in bonnet is the simplest bonnet, offering a durable, pressure-tight seal. Union bonnet is suitable for applications requiring frequent inspection or cleaning. It also gives the body added strength. A bonnet attached with bolts is used for larger or higher pressure applications. The bonnet also contains the packing, a wearable material that maintains the seal between the bonnet and the stem during valve cycling.

Plug or Disc (Disk)

The closure member of the valve, plugs are connected to the stem which is slid or screwed up or down to throttle the flow. Plugs are typically of the balance or unbalanced type. Unbalanced plugs are solid and are used with smaller valves or with low pressure drops across the valve.

The advantages are simpler design, with one possible leak path at the seat and usually lower cost. The disadvantages are the limited size; with a large unbalanced plug the forces needed to seat and hold the flow often becomes impractical. Balanced plugs have holes through the plug. Advantages include easier shut off as the plug does not have to overcome static forces. However, a second leak path is created between the plug and the cage, and cost is generally higher.

Stem

The stem serves as a connector from the actuator to the inside of the valve and transmits this actuation force. Stems are either smooth for actuator controlled valves or threaded for manual valves. The smooth stems are surrounded by packing material to prevent leaking material from the valve. This packing is a wearable material and will have to be replaced during maintenance. With a smooth stem the ends are threaded to allow connection to the plug and the actuator. The stem must not only withstand a large amount of compression force during valve closure, but also have high tensile strength during valve opening. In addition, the stem must be very straight, or have low run out, in order to ensure good valve closure. This minimum run out also minimises wear of the packing contained in the bonnet, which provides the seal against leakage. The stem may be provided with a shroud over the packing nut to prevent foreign bodies entering the packing material, which would accelerate wear.

Cage

The cage is part of the valve that surrounds the plug and is located inside the body of the valve. Typically, the cage is one of the greatest determiners of flow within the valve. As the plug is moved more of the openings in the cage are exposed and flow is increased and vice versa. The design and layout of the openings can have a large effect on flow of material (the flow characteristics of different materials at temperatures, pressures that are in a range). Cages are also used to guide the plug to the seat of the valve for a good shutoff, substituting the guiding from the bonnet.

Seat Ring

The seat ring provides a stable, uniform and replaceable shut off surface. Seat rings are usually held in place by pressure from the fastening of the bonnet to the top of the body. This pushes the cage down on the lip of the seat ring and holds it firmly to the body of the valve. Seat rings may also be threaded and screwed into a thread cut in the same area of the body. However this method makes removal of the seat ring during maintenance difficult if not impossible. Seat rings are also typically beveled at the seating surface to allow for some guiding during the final stages of closing the valve.

Economical globe valves or stop valves with a similar mechanism used in plumbing often have a rubber washer at the bottom of the disc for the seating surface, so that rubber can be compressed against the seat to form a leak-tight seal when shut.

Materials

Typically globe valves are made of metallic alloys, although some synthetic materials are available. These materials are chosen based on pressure, temperature, controlled media properties. Corrosive and/ or erosive process streams may require a compromise in material selection or exotic alloys or body coatings to minimise these material interactions and extend the life of the valve or valve trim components. Typically, carbon steel alloys are specified for noncorrosive applications. Other alloys such as Hastelloy, Monel, Inconel and others are available.

Packing material must also be considered during valve selection. Typically the requirement for a low friction packing conflict with a durable material that will provide low maintenance requirements during service life. Corrosive applications can further complicate packing material selection as the typical packing materials may or may not be compatible with the processed materials. Typically graphite or PTFE is used due to its low friction coefficient. Enviro-seal applications also have the availability of constant applied force (live-load) packing. While more complex, it allows for constant packing force load throughout the life of the packing material. This packing helps meet contemporary environmental laws.

Needle Valve

A needle valve is a type of valve having a small port and a threaded, needle-shaped plunger. It allows precise regulation of flow, although it is generally only capable of relatively low flow rates.

Construction and Operation

A needle valve has a relatively small orifice with a long, tapered seat, and a needle-shaped plunger, on the end of a screw, which exactly fits this seat. As the screw is turned and the plunger retracted, flow between the seat and the plunger is possible; however, until the plunger is completely retracted the fluid flow is significantly impeded. Since it takes many turns of the fine-threaded screw to retract the plunger, precise regulation of the flow rate is possible. The virtue of the needle valve is from the vernier effect of the ratio between the needle's length and its diametre, or the difference in diametre between needle and seat. A long travel axially (the control input) makes for a very small and precise change radially (affecting the resultant flow). Needle valves may be used in vacuum systems, when a very precise control of gas flow is required, at low pressure, such as when filling gas-filled vacuum tubes, gas lasers and similar devices.

Uses

Needle valves are usually used in flow metring applications, especially when a constant, calibrated, low flow rate must be maintained for some time, such as the idle fuel flow in a carburetor.

Note that the float valve of a carburetor (controlling the fuel level within the carburetor) is *not* a needle valve, although it is commonly described as one. It uses a bluntly conical needle, but it seats against a square-edged seat rather than a matching cone. The intention here is to obtain a well-defined seat between two narrow mating surfaces, giving firm shutoff of the flow from only a light float pressure.

Since flow rates are low and many turns of the valve stem are required to completely open or close, needle valves are not used for simple shutoff applications. Since the orifice is small and the force advantage of the fine-threaded stem is high, needle valves are usually easy to shut off completely, with merely "finger tight" pressure. Small, simple needle valves are often used as bleed valves in hot water heating applications. Unlike a ball valve, or valves with a rising stem, it is not easy to tell from examining the handle position whether the valve is open or closed.

Pinch Valve

A pinch valve is a full bore or fully ported type of control valve which uses a pinching effect to obstruct fluid flow. There are a few types of pinch valves based upon application.

Fluids: Pinch valves used for fluids usually employ a device that directly contacts process tubing. Forcing the tubing together will create a seal that is equivalent to the tubing's permeability.

Particulates

Major components of a pinch valve consists of body and a sleeve. The sleeve will contain the flow media and isolate it from the environment hence reducing contamination to the environment. Generally used for slurries or processes with entrained solids, because the flexible rubber sleeve allows the valve to close droptight around solids—solids that would typically be trapped by the seat or stuck in crevices in globe, diaphragm, butterfly, gate, or ball valves. The sleeve material can be selected upon the corrosiveness and abrasiveness of the flow media, a suitable synthetic polymer can be chosen. A pinch valve may be the best type of valve for flow control application if the operation temperature is within the limit of the polymer.

Air Operated Pinch Valves

Air operated Pinch Valves are very easily constructed. They generally only consist of an extremely elasticised reinforced rubber hose, a type of housing, and two socket end covers or flanges. The rubber hoses of the air operated Pinch Valves are usually press-fitted and centred into the housing ends by the socket end covers or flanges.

Function of Air Operated Pinch Valves

The air operated Pinch Valve works without any additional actuator; all it needs to close or operate is 30 psi air supply into the Pinch Valve body. As soon as the air supply becomes interrupted and the volume of air exhausts, the elastic rubber hose starts to open due to its great impact resilience.

Advantage of Air Operated Pinch Valves

The most important benefits of using air operated Pinch Valves are their complete and true full bore, and the 100% tight shut off – even on solids such as granules, powders, pellets, chippings, fibres, slivers, any kind of slurries and many more aggressive products.

Usually conventional valves such as ball valves, butterfly valves, piston valves or gate valves tend to fail when in process with the above mentioned aggressive products. The reason is that the valve body seat or the gate/piston of the valves wears out too quickly and the valve can no longer shut off tightly, whereas air operated Pinch Valves work without almost any wear of the elastic rubber hose because the kinetic energy of the solids are absorbed through the extremely high elasticity of the rubber. The rubber mixture is also anti abrasive.

Industrial Sectors

Therefore using air operated Pinch Valves in aggressive product processes saves a lot down time, and guarantees a free flow through the valve, avoiding any blockages.

Several rubber qualities are available for air operated Pinch Valves such as natural rubber, EPDM, nitrile, viton, neoprene and butyl. Different housings and end covers/flange materials such as aluminium, plastics and stainless steel are also available; resulting in air operated Pinch Valves being commonly used in a wide range of many different industries such as:

- Bulk and solids handling,
- Cement industry,
- Waste water industry,

- Chemical industry,
- Food industry,
- Beverage industry,
- Ceramic-/Glass-/Plastic industry.

Plus Many More

On a smaller scale, pinch valves are used in some water pistols, notably the original Super Soaker 50.

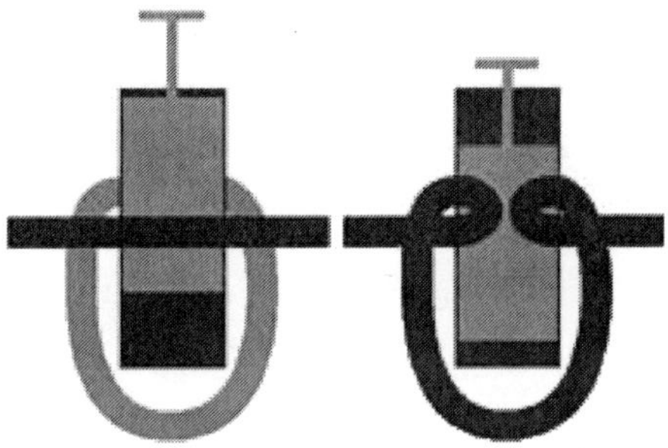

Figure: Piston valve in a brass instrument

A 'piston valve' is a device used to control the motion of a fluid along a tube or pipe by means of the linear motion of a piston within a chamber or cylinder.

Examples of piston valves are:

- The valves used in many brass instruments
- The valves used in pneumatic cannons
- The valves used in many stationary steam engines and steam locomotives

Brass Instruments

Cylindrical piston valves called Perinet valves (after their inventor, François Perinet), are used to change the length of tube in the playing of most brass instruments, particularly the trumpet-like members of the family (cornet, flugelhorn, saxhorn, etc.).

Other brass instruments use rotary valves. Notably the orchestral horns and many tuba models, but also a number of rotary-valved variants of those brass instruments which more commonly employ piston valves. The first piston-valved musical instruments were developed just after the start of the 19th century. The Stölzel valve (invented by Heinrich Stölzel in 1814) was an early variety. In the mid 19th century the Vienna valve was an improved design. However most professional musicians preferred rotary valves for quicker, more reliable action, until better designs of piston valves were mass manufactured in towards the end of the 19th century.

Pneumatic Cannons

A piston valve can also refer to a 2-way 2-position, pilot-operated spool valve. The term is extremely popular among spud gun enthusiasts who often build homemade piston valves for use in pneumatic cannons. Valves are typically constructed primarily from pipe fittings and machined plastics or metals. The inside of a piston valve contains a piston that blocks the output when the valve is pressurized, and a volume of air behind the piston. When the pressure behind the piston is released the piston is pushed back by the force of the pressure from the input. This allows the valve to be opened by a much smaller pilot valve, with speeds faster than possible with just a manually operated valve. Functionally these types of valves are comparable to quick exhaust valves. This type of piston valve is also sometimes referred to as a back-pressure valve.

Piston Valve (Steam Engine)

Piston valves are one form of valve used to control the flow of steam within a steam engine or locomotive. They control the admission of steam into the cylinders, and its exhaustion from the cylinders after use, enabling a locomotive to move under its own power.

Overview: In the 19th century, steam locomotives used slide valves to control the flow of steam into and out of the cylinders. In the 20th century, slide valves were gradually superseded by piston valves, particularly in engines using superheated steam.

There were two reasons for this:

- It is difficult to lubricate slide valves adequately in the presence of superheated steam.
- With piston valves, the steam passages can be made shorter. This, particularly following the work of André Chapelon, reduces resistance to the flow of steam and improves efficiency.

The usual locomotive valve gears such as Stephenson, Walschaerts, and Baker valve gear, can be used with either slide valves or piston valves. Where poppet valves are used, a different gear, such as Caprotti valve gear is needed.

Examples

The Swannington incline winding engine of 1833 incorporated a piston valve.

The Swannington incline winding engine on the Leicester and Swannington Railway, manufactured by The Horsely Coal & Iron Company in 1833, shows a very early use of the piston valve. Piston valves had been used a year or two previously in the horizontal engines manufactured by Taylor and Martineau of London, but did not become general for stationary or locomotive engines until the end of the 19th century.

Design Principles

When on the move, a steam locomotive requires steam to enter the piston at a controlled rate, in order for the locomotive to function efficiently. This entails controlling the admission and exhaustion of steam to and from the cylinders. Steam enters and leaves the valve through a steam port, which is located in the middle position of the piston valve. Where the valve is in contact with the steam ports, a consideration of the "lap" and "lead" is required.

Lap

The "Lap" is the amount by which the valve overlaps each steam port at the middle position of each valve. However, there are two different types of "Lap."

The first kind is the "steam lap," which is the amount by which the valve overlaps the port on the live steam side of the cylinder. Secondly, there is the "exhaust lap," which is the amount by which the valve overlaps the port on the exhaust side of the cylinder. "Exhaust lap" is generally given to slow-running locomotives. This is because it allows the steam to remain in the cylinder for the longest possible amount of time before being expended as exhaust, therefore increasing efficiency. Shunter locomotives tended to be equipped with this addition.

The "Negative exhaust lap", also commonly termed "exhaust clearance," is the amount the port is open to exhaust when the valve is in mid-position, and this is used on many fast-running locomotives to give a free exhaust. The amount seldom exceeds 1/16 in. when exhaust clearance is given; the cylinder on both sides of the piston is open to exhaust at the same time when the valve is passing through the mid-position, which is only momentary when running.

Lead

The "lead" of the valve is the amount by which the steam port is open when the piston is static at front or back dead centre. Pre-admission of steam fills the clearance space between the cylinder and piston and ensures maximum cylinder pressure at the commencement of the stroke. "Lead" is particularly necessary on locomotives designed for high speeds, under which conditions the valve events are taking place in rapid succession.

Valve Travel

Long-travel piston valves allow the use of large steam ports to ease the flow of steam in to, and out of, the cylinder.

Slide Valve

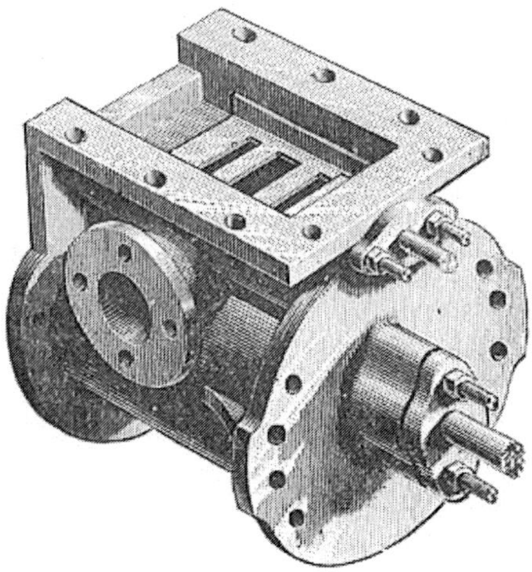

Figure: Cylinder, with slide valve removed to show ports

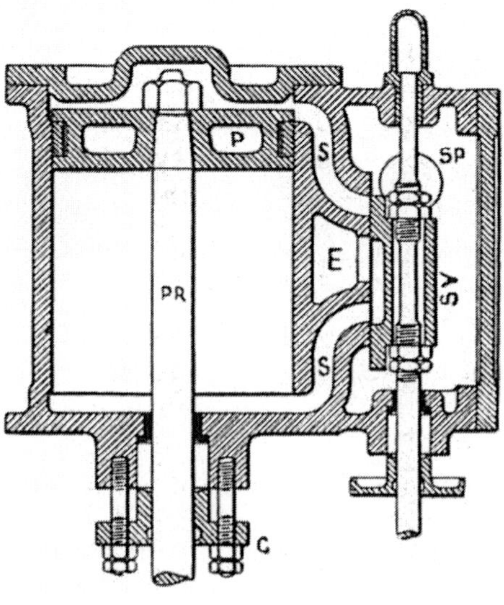

Figure: A double-acting slide valve cylinder.

Steam enters via the steam port *SP*, and is admitted by the slide valve *SV* through the upper passage *S* to push down the piston *P*. At the same time, exhaust steam from below the piston passes back up the lower passage *S*, via the valve cavity, to exhaust *E*. As the piston descends, the valve moves upwards to admit steam below the piston and release exhaust from above. The slide valve is a rectilinear valve used to control the admission of steam into, and emission of exhaust from, the cylinder of a steam engine.

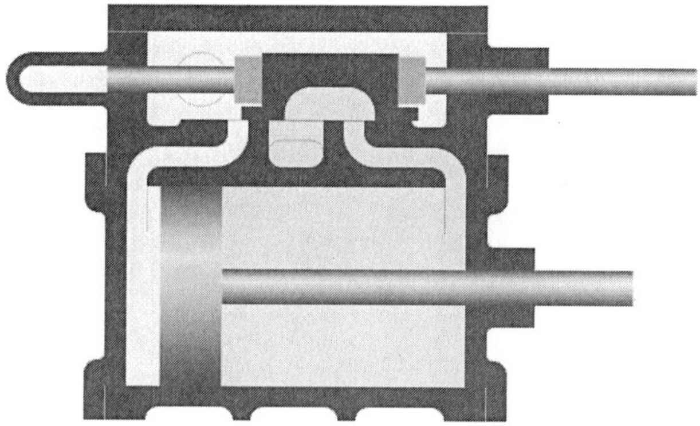

Figure: The image description page explains how this SVG animation can be viewed

Use

In the 19th century, most steam locomotives used slide valves to control the flow of steam into and out of the cylinders. In the 20th century, slide valves were gradually superseded by piston valves, particularly in engines using superheated steam.

There were two reasons for this:

- With piston valves, the steam passages can be made shorter. This reduces resistance to the flow of steam and improves efficiency.
- It is difficult to lubricate slide valves adequately in the presence of superheated steam.

Murdoch's D slide valve

The D slide valve, or more specifically Long D slide valve, is a form of slide valve, invented by William Murdoch and patented in 1799. It is named after the hollow central D-sectioned piston.

This valve worked by "connecting the upper and lower valves so as to be worked by one rod or spindle, and in making the stem or tube which connects them hollow, so as to serve for an induction pipe to the upper end of the cylinder." This allowed two valves to do the work of four.

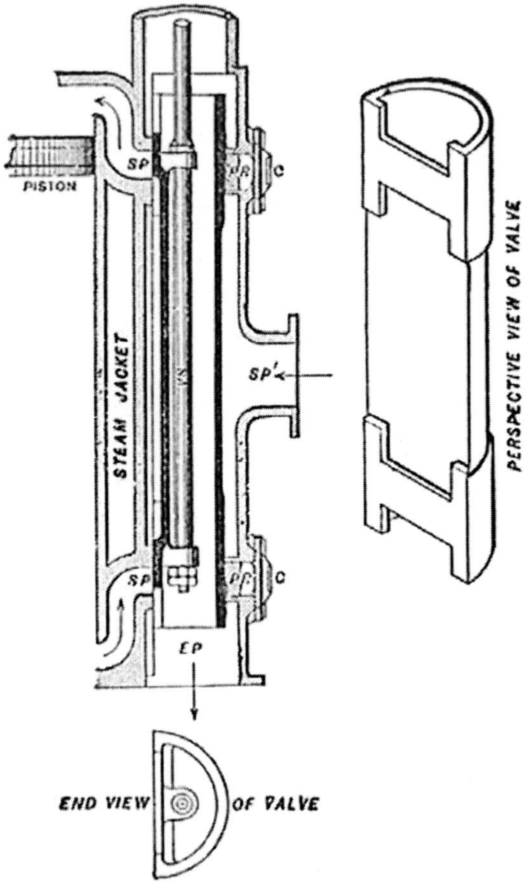

Figure: Murdoch's long D slide valve

The above description (referring to upper and lower valves) clearly relates to an engine with a vertical cylinder, such as a beam engine. Where the cylinders are horizontal, as in a steam locomotive, the valves would be side-by-side.

Balanced Slide Valve

The balanced slide valve was invented by the Scottish engineer Alexander Allan. It was not much used in the UK but, at one time, had great popularity in the USA. It gave some of the advantages of a piston valve to a slide valve by relieving the pressure on the back of the valve, thus reducing friction and wear.

Plug Valve

Plug valves are valves with cylindrical or conically-tapered "plugs" which can be rotated inside the valve body to control flow through the valve. The plugs in plug valves have one or more hollow passageways going sideways through the plug, so that fluid can flow through the plug when the valve is open. Plug valves are simple and often economical.

When the plug is conically-tapered, the stem/handle is typically attached to the larger diametre end of the plug. Plug valves usually do not have bonnets but often have the end of the plug with the handle exposed or mostly exposed to the outside. In cases like that, there is usually not much of a stem. The stem and handle often come in one piece, often a simple, approximately L-shaped handle attached to the end of the plug. The other end of the plug is often exposed to the outside of the valve too, but with a mechanism which retains the plug in the body.

The simplest and most common general type of plug valve is a 2-port valve, which has two positions, *open* to allow flow, and *shut* (closed) to stop flow. Ports are openings in the valve body through which fluid can enter or leave. The plug in this kind of valve has one passageway going through it. The ports are typically at opposite ends of the body; therefore, the plug is rotated a fourth of a full turn to change from open to shut positions. This makes this kind of plug valve a *quarter-turn valve.* There is often a mechanism limiting motion of the handle to a quarter turn, but not in glass stopcocks.

Slightly conically-tapered metal (often brass) plug valves are often used as simple shut-off valves in household natural gas lines.

It is also possible for a plug valve to have more than two ports. In a 3-way plug valve, flow from one port could be directed to either the second or third port. A 3-way plug valve could also be designed to shift flow between ports 1 and 2, 2 and 3, or 1 and 3, and possibly even connect all three ports together. The flow-directing possibilities in multi-port plug valves are similar to the possibilities in corresponding multi-port ball valves or corresponding multi-port valves with a rotor. An additional possibility in plug valves is the have one port on one side of the plug valve and two ports on the other side, with two diagonal and parallel fluid pathways inside the plug. In this case the plug can be rotated 180° to connect the port on the one side to either of the two ports on the other side. Stopcocks used in laboratory glassware are typically forms of conically-tapered plug valves. When fused with the glassware, the valve bodies are made of glass. Otherwise, they can be made of an inert plastic such as Teflon. The plugs can be made of a similar plastic or glass. When the plug is made of glass, the handle and plug are fused together in one piece out of glass. When glass is used for both the stopcock body and the plug, the contacting surfaces between them are special ground glass surfaces often with stopcock grease in between. Special glass stopcocks are made for vacuum applications, such as in use with vacuum manifolds. Stopcock grease is always used in high vacuum applications to make the stopcock air-tight. Also if the plug valve is "locked" from being in the open or closed position for an extended amount of time lubricant can be added through the greaser with the valve in service.

Poppet Valve

Poppet valve & components, including valve stem oil seal. The yellow paint is a spring strength marking

A poppet valve (also called mushroom valve) is a valve typically used to control the timing and quantity of gas or vapour flow into an engine. It consists of a hole, usually round or oval, and a tapered plug, usually a disk shape on the end of a shaft also called a valve stem. The shaft guides the plug portion by sliding through a valve guide. In most applications a pressure differential helps to seal the valve and in some applications also open it. Poppet valves date from at least the 1770s, when Watt used them on his beam engines.

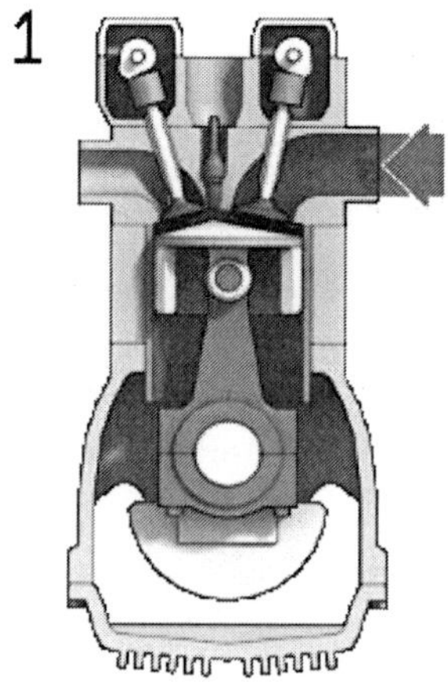

Figure: Poppet valves in action at the top of the cylinder

Operation: The operating principle of poppet valves is described in the online article "How Poppet Valves Work". In most cases it is beneficial to have a "balanced poppet" in a direct-acting valve. Less force is needed to move the poppet because all forces on the poppet are nullified by equal and opposite forces. The solenoid coil has to counteract only the spring force.

Applications

Poppet valves are used in many industrial processes, from controlling the flow of milk to isolating sterile air in the semiconductor industry. However, they are most well known for their use in internal combustion and steam engines, as described below.

Presta and Schrader valves used on pneumatic tires are examples of poppet valves. The Presta valve has no spring and relies on a pressure differential for opening and closing while being inflated.

Internal Combustion Engine

Poppet valves are used in most piston engines to open and close the intake and exhaust ports in the cylinder head. The valve is usually a flat disk of metal with a long rod known as the valve stem attached to one side.

The stem is used to push down on the valve and open it, with a spring generally used to return it to the closed position when the stem is not being depressed. At high revolutions per minute (RPM), the inertia of the spring makes it too slow to return the valve to its seat between cycles, leading to 'valve float'. In this situation desmodromic valves are used which, being closed by a positive mechanical action instead of by a spring, are able to cycle at the high speeds required in, for instance, motorcycle and auto racing engines.

The engine normally operates the valves by pushing on the stems with cams and cam followers. The shape and position of the cam determines the valve lift and when and how quickly (or slowly) the valve is opened. The cams are normally placed on a fixed camshaft which is then geared to the crankshaft, running at half crankshaft speed in a four-stroke engine. On high-performance engines, the camshaft is movable and the cams have a varying height, so by axially moving the camshaft in relation with the engine RPM, also the valve lift varies.

For certain applications the valve stem and disk are made of different steel alloys, or the valve stems may be hollow and filled with sodium to improve heat transport and transfer. Although better heat conductors, aluminium cylinder heads require steel valve seat inserts, while cast-iron cylinder heads often used integral valve seats in the past. Because the valve stem extends into lubrication in the cam chamber, it must be sealed against blow-by to prevent cylinder gases from escaping into the crankcase, even though the stem to valve clearance is very small, typically 0.04-0.06 mm. A rubber lip-type seal ensures that excessive amounts of oil are not drawn in from the crankcase on the induction stroke and that exhaust gas does not enter the crankcase on the exhaust stroke. Worn valve guides or defective oil seals are characterised by a puff of blue smoke from the exhaust when pressing back down on the accelerator pedal after allowing the engine to over-run, such as when changing gears.

Valve Position

In very early engine designs the valves were 'upside down' in the block, parallel to the cylinders - the so called L-head engine because of the shape of the cylinder and combustion chamber, also called 'flathead engine' as the top of the cylinder head is flat. Although this design makes for simplified and cheap construction, it has two major drawbacks; the tortuous path followed by the intake charge limits air flow and effectively prevents speeds greater than 2,000–2,500 RPM, and the travels of the exhaust through the block can cause overheating under sustained heavy load. This design evolved into 'Intake Over Exhaust', IOE or F-head, where the intake valve was in the head and the exhaust valve was in the block; later both valves moved to the head.

In most such designs the camshaft remained relatively near the crankshaft, and the valves were operated through pushrods and rocker arms. This led to significant energy losses in the engine, but was simpler, especially in a V engine where one camshaft can actuate the valves for both cylinder banks; for this reason, pushrod engine designs persisted longer in these configurations than others. More modern designs have the camshaft on top of the cylinder head, pushing directly on the valve stem (again through cam followers, also known as tappets), a system known as *overhead camshaft*; if there is just one camshaft, this is a single overhead cam or *SOHC* engine. Often there are two camshafts, one for the intake and one for exhaust valves, creating the dual overhead cam, or *DOHC*. The camshaft is driven by the crankshaft - through gears, a chain or a timing belt.

Valve Wear

In the early days of engine building, the poppet valve was a major problem. Metallurgy was not what it is today, and the rapid opening and closing of the valves against the cylinder heads led to rapid wear.

They would need to be re-ground every two years or so, in an expensive and time-consuming process known as a *valve job*. Adding tetra-ethyl lead to the petrol reduced this problem to some degree, as the lead would coat the valve seats, in effect lubricating the metal. Valve seats made of improved alloys such as stellite have generally made this problem disappear completely and made leaded fuel unnecessary.

Valve burn (overheating) is another major problem. It causes excessive valve wear and defective sealing, as well as engine knocking. It can be solved by valve cooling systems that use water or oil as a coolant. In high performance engines sometimes sodium-cooled valve stems are used. These hollow valve stems are partially filled with sodium and act as a heat pipe.

Steam Engine

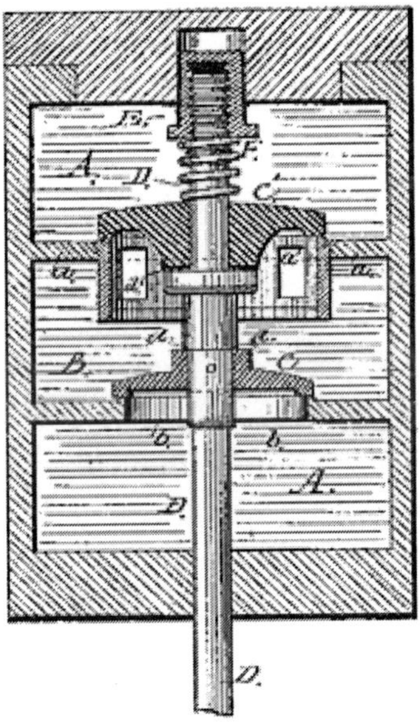

Figure: Balanced Poppet Valve from U.S. Patent 339,809. High pressure steam enters at A and exits at B. The valve stem D moves up to open the valve discs C

James Watt was using poppet valves to control the flow of steam into the cylinders of his beam engines in the 1770s. A sectional illustration of Watt's beam engine of 1774 using the device is found in Thurston 1878:98, and Lardner (1840) provides an illustrated description of Watt's use of the poppet valve.

When used in high-pressure applications, for example, as admission valves on steam engines, the same pressure that helps seal poppet valves also contributes significantly to the force required to open them. This has led to the development of the balanced poppet or double beat valve, in which two valve plugs ride on a common stem, with the pressure on one plug largely balancing the pressure on the other. In these valves, the force needed to open the valve is determined by the pressure and the difference between the areas of the two valve openings. Sickels patented a valve gear for double-beat poppet valves in 1842. Criticism was reported in the journal Science in 1889 of equilibrium poppet valves (called by the article the 'double or balanced or American puppet valve') in use for paddle steamer engines, that by its nature it must leak 15 percent.

Poppet valves have been used on steam locomotives, often in conjunction with Lentz or Caprotti valve gear.

British examples include:

- LNER Class B12
- LNER Class D49
- LNER Class P2
- LMS Stanier Class 5 4-6-0
- BR standard class 5
- BR standard class 8 71000 Duke of Gloucester.

Sentinel Waggon Works used poppet valves in their steam wagons and steam locomotives. Reversing was achieved by a simple sliding camshaft system.

The poppet valve was also used on the American Pennsylvania Railroad's T1 duplex locomotives, although the valves commonly failed due to the fact that the locomotives were commonly operated in excess of 160 km/h (100 mph), and the valves were not meant for the stresses of such speeds.

Relief Valve

The relief valve (RV) is a type of valve used to control or limit the pressure in a system or vessel which can build up by a process upset, instrument or equipment failure, or fire.

The pressure is relieved by allowing the pressurised fluid to flow from an auxiliary passage out of the system. The relief valve is designed or set to open at a predetermined set pressure to protect pressure vessels and other equipment from being subjected to pressures that exceed their design limits. When the set pressure is exceeded, the relief valve becomes the "path of least resistance" as the valve is forced open and a portion of the fluid is diverted through the auxiliary route. The diverted fluid (liquid, gas or liquid–gas mixture) is usually routed through a piping system known as a *flare header* or *relief header* to a central, elevated gas flare where it is usually burned and the resulting combustion gases are released to the atmosphere. As the fluid is diverted, the pressure inside the vessel will drop. Once it reaches the valve's reseating pressure, the valve will close. The *blowdown* is usually stated as a percentage of set pressure and refers to how much the pressure needs to drop before the valve reseats. The blowdown can vary from roughly 2–20%, and some valves have adjustable blowdowns. In high-pressure gas systems, it is recommended that the outlet of the relief valve is in the open air. In systems where the outlet is connected to piping, the opening of a relief valve will give a pressure build up in the piping system downstream of the relief valve. This often means that the relief valve will not re-seat once the set pressure is reached. For these systems often so called "differential" relief valves are used.

This means that the pressure is only working on an area that is much smaller than the openings area of the valve. If the valve is opened the pressure has to decrease enormously before the valve closes and also the outlet pressure of the valve can easily keep the valve open. Another consideration is that if other relief valves are connected to the outlet pipe system, they may open as the pressure in exhaust pipe system increases. This may cause undesired operation.

In some cases, a so-called *bypass valve* acts as a relief valve by being used to return all or part of the fluid discharged by a pump or gas compressor back to either a storage reservoir or the inlet of the pump or gas compressor. This is done to protect the pump or gas compressor and any associated equipment from excessive pressure. The bypass valve and bypass path can be internal (an integral part of the pump or compressor) or external (installed as a component in the fluid path). Many fire engines have such relief valves to prevent the overpressurization of fire hoses.

In other cases, equipment must be protected against being subjected to an internal vacuum (i.e., low pressure) that is lower than the equipment can withstand. In such cases, *vacuum relief valves* are used to open at a predetermined low pressure limit and to admit air or an inert gas into the equipment so as control the amount of vacuum.

Technical Terms: In the petroleum refining, petrochemical and chemical manufacturing, natural gas processing and power generation industries, the term relief valve is associated with the terms pressure relief valve (PRV), pressure safety valve (PSV) and safety valve.

In practice, people often do not stick to the technical distinctions between the most common abbreviations: SRV, PRV, SV and RV

Pressure relief valve (PRV) or pressure safety valve (PSV). The difference being that PSVs have a manual lever to activate the valve in case of emergency. Most PRV are spring operated. At lower pressures some use a diaphragm in place of a spring. The oldest PRV designs use a weight to seal the valve.

Set pressure: When increasing system pressure reaches this value the PRV opens. Accuracy of set pressure often follows guidelines set by the ASME.

Relief valve (RV): A valve used on a liquid service, which opens proportionally as the increasing pressure overcomes the spring pressure.

Safety valve (SV): Used in gas service. Most SV are full lift or snap acting, they pop open all the way.

Safety relief valve (SRV): A PRV that can be used for gas or liquid service. But set pressure will usually only be accurate for one type of fluid at a time (the type it was set with).

Pilot-operated relief valve (POSRV, PORV, POPRV): device that relieves by remote command from a pilot valve that is connected to the upstream system pressure. Low-pressure safety valve (LPSV): automatic system that relieves by static pressure on a gas. The pressure is small and near the atmospheric pressure.

Vacuum pressure safety valve (VPSV): automatic system that relieves by static pressure on a gas. The pressure is small, negative and near the atmospheric pressure.

Low and vacuum pressure safety valve (LVPSV): automatic system that relieves by static pressure on a gas. The pressure is small, negative or positive and near the atmospheric pressure.

Snap acting: The opposite of modulating, refers to a valve that "pops" open, it goes into full lift in milliseconds. Usually accomplished with a skirt on the disc so that the fluid passing the seat suddenly affects a larger area and creates more lifting force.

Modulating: Opens in proportion to the overpressure formed.

Legal and Code Requirements in Industry

The main standards, laws or directives are:

- ASME (American Society of Mechanical Engineers) Boiler & Pressure Vessel Code, Section VIII Division 1 and Section I
- API (American Petroleum Institute) Recommended Practice 520/521, API Standard 2000 et API Standard 526
- ISO 4126 (International Organisation for Standardisation)
- EN 764-7 (European Standard based on pressure Equipment Directive 97/23/EC)
- AD Merkblatt (German)
- PED 97/23/EC (Pressure Equipment Directive – European Union).

Diers

Formed in 1976, the Design Institute for Emergency Relief Systems was a consortium of 29 companies under the auspices of the American Institute of Chemical Engineers (AIChE) that developed methods for the design of emergency relief systems to handle runaway reactions. Its purpose was to develop the technology and methods needed for sizing pressure relief systems for chemical reactors, particularly those in which exothermic reactions are carried out. Such reactions include many classes of industrially important processes including polymerizations, nitrations, diazotizations, sulphonations, epoxidations, aminations, esterifications, neutralizations and many others. Pressure relief systems can be difficult to design, not least because what is expelled can be gas/vapour, liquid, or a mixture of the two – just as with a can of carbonated drink when it is suddenly opened.

DIERS investigated the two-phase vapour–liquid onset / disengagement dynamics and the hydrodynamics of emergency relief systems with extensive experimental and analysis work. Of particular interest to DIERS were the prediction of two-phase flow venting and the applicability of various sizing methods for two-phase vapour-liquid flashing flow. DIERS became a user's group in 1985.

European DIERS Users' Group (EDUG) is a group of mainly European industrialists, consultants and academics who use the DIERS technology. The EDUG started in the late 1980s and has an annual meeting. A summary of many of key aspects of the DIERS technology has been published in the UK by the HSE.

Reed Valve

Reed valves are a type of check valve which restrict the flow of fluids to a single direction, opening and closing under changing pressure on each face. Modern versions often consist of flexible metal or composite materials (fibreglass or carbon fibre).

Applications

Traditional: Reed-valves, normally a leather flap covering a hole, are amongst the earliest form of automatic flow control for liquids and gases. They have been used for thousands of years in water pumps and for hundreds of years in bellows for high-temperature forges and musical instruments such as church organs and accordions. In nature, heart valves operate in a somewhat similar fashion.

Pumps

Reed valves are used in some reciprocating compressor designs, and in the pumping element of some musical instruments, large and small.

Two-Stroke Engines

Reed valves are commonly used in high-performance versions of the two-stroke engine, where they control the fuel-air mixture admitted to the cylinder. As the piston rises in the cylinder a vacuum is created in the crankcase beneath the piston. This vacuum opens the valve and admits the fuel-air mixture into the crankcase. As the piston descends, it raises the crankcase pressure causing the valve to close to retain the mixture and pressurize it for its eventual transfer through to the combustion chamber. Composite materials are preferred in racing engines, especially in kart racing, because the stiffness of the petals can be easily tuned and they are relatively safe in failure. High-speed impact takes its toll on all reed valves, with metal valves suffering in fatigue. The physical inertia of reed valves means that they are not as entirely precise in action as rotary valves, a rotary valve engine may run better than a reed valve engine at a small rpm range but the reed valve engine often runs better over a wider rpm range. More sophisticated designs partly address this by creating multi-stage reeds with smaller, more responsive reeds within larger ones that provide more volume later in the cycle. Nevertheless, current technology favours reed valves almost to the exclusion of rotary valves due to their simplicity and low implementation costs.

Pulse Jets

Reed valves are used in the cheap but inefficient pulse jet engine, such as the one used by the Argus engine in the German V-1 (flying bomb). The valves at the front of the cylindrical engine are opened by the low pressure in the combustion chamber caused by the resonance of the air column in the engine, fuel is squirted into the combustion chamber and ignited by the hot combustion gases of the previous cycle. Once the charge has expanded and mostly left the engine, pressure inside drops again to below-atmospheric values and the reed valve allows fresh air to enter and the cycle be repeated. Some ram-air pressure due to forward motion helps scavenging and filling the combustion chamber with the new, fresh air charge, thus improving the power of the engine at higher speeds.

Design and Modelling

Reed valves are designed considering the pressure gradient and mass flow. The pressure gradient is used to evaluate the valve lift during open condition; the lift and overall component geometry (considering also a pressure loss coefficient) are then used to calculate the mass flow.

For high speed applications (compressors and engines) the dynamic response has to be considered. A simple approach consists in the evaluation of first eigenvalue that is compared with exciting frequency. Design of reed valves can be refined using simulations. The dynamic of petals can be studied neglecting the coupling between fluid and structure: in this case the evolution of the structural part are simulated using lumped parameters models or FEM models, discharge coefficients at various valve lift are evaluated with experiments or CFD simulations. The study of the complete system needs an integrated Fluid-structure interaction model.

Rotary Valve

A rotary valve is a type of valve in which the rotation of a passage or passages in a transverse plug regulates the flow of liquid or gas through the attached pipes. The common stopcock is the simplest form of rotary valve.

Rotary valves have been applied in numerous applications, including:

- Changing the pitch of brass instruments.
- Controlling the steam and exhaust ports of steam engines, most notably in the Corliss steam engine.
- Periodically reversing the flow of air and fuel across the open hearth furnace.
- Loading sample on chromatography columns.
- Certain types of two-stroke engines.

Use in Brass Instruments

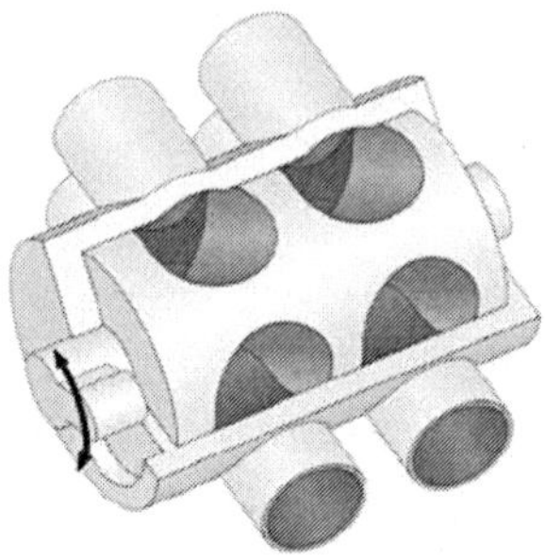

Figure: Rotary valve for a double French horn

In the context of brass instruments, rotary valves are found on horns, trumpets, trombones, flugelhorns, and tubas. Many European trumpet players tend to favour rotary valves.

Trombone F-attachment valves are usually rotary, with several variations on the basic design also in use, such as the Thayer axial-flow valve and Hagmann valve.

Joseph Riedlin is credited with the first use of rotary valves on brass instruments in 1832.

Use in Industry

In industry, a rotary valve (which can also be called airlock) is often used to enter or extract material from two chambers with different pressure level.

As part of the material exchange process, the valve is often used as a measuring or metring device.

Use in Engine Design

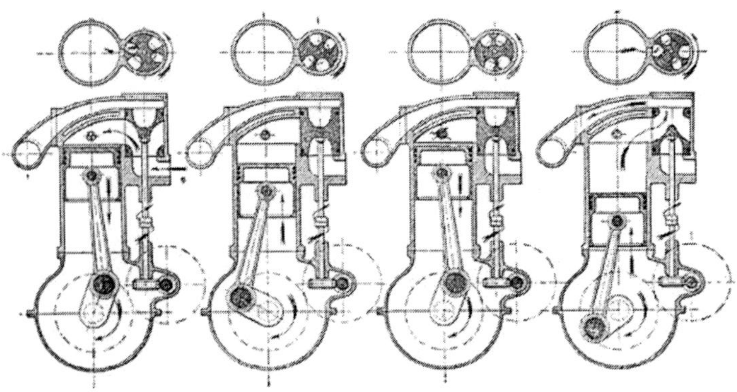

Figure: Itala rotary valve engine (1912).

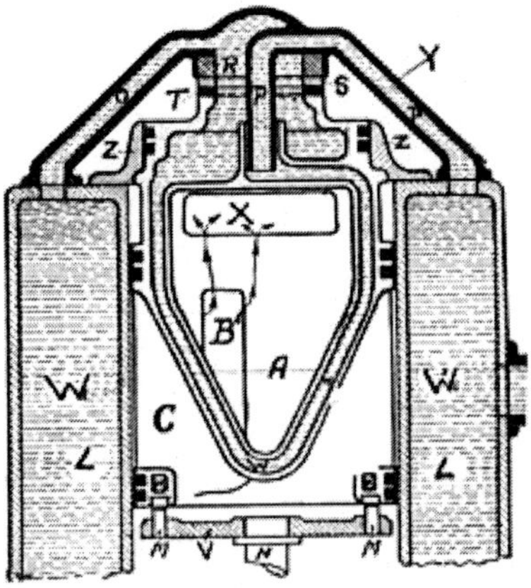

Figure: Itala rotary valve cooling (1919).

The rotary valve combustion engine possesses several significant advantages over the conventional assemblies, including significantly higher compression ratios and rpm, meaning more power, a much more compact and light-weight cylinder head, and reduced complexity, meaning higher reliability and lower cost. As inlet and exhaust are usually combined special attention should be given to valve cooling to avoid engine knocking.

Rotary valves have been used in several different engine designs. In Britain, the *National Engine Company Ltd* advertised its rotary valve engine for use in early aircraft, at a time when poppet valves were prone to failure by sticking or burning.

From the 1930s, Frank Aspin developed a design with a rotary valve that rotated on the same axis as the cylinder bore, but with limited success. Kawasaki and others have also used rotary valves in two-stroke motorcycle engines, where the arrangement helps to prevent reverse flow back into the intake port during the compression stroke.

Austrian engine manufacturer Rotax used rotary intake valves in their now out-of-production 64 hp (48 kW) Rotax 532 two-stroke engine design and continues to use rotary intake valves in the 532's successor, the current-production 64 hp (48 kW) Rotax 582.

US company *Coates International Ltd* has developed a spherical rotary valve for internal combustion engines which replaces the poppet valve system. This particular design is four-stroke, with the rotary valves operated by overhead shafts in lieu of overhead camshafts (i.e. in line with a bank of cylinders). The first sale of such an engine was part of a natural gas engine-generator.

UK company *RCV Engines Ltd* uses rotating cylinder liner technology as a specialised form of rotary valve in their model engine and small-engine line-up.

Rotary valves are highly suitable for high reving engines, such as those used in sportscars and F1 racing cars, on which traditional poppet valves with springs can fail due to spring resonance and where the desmodromic valve gear is too heavy, large in size and too complex to time and design properly. Rotary valves allow for a more compact and lightweight cylinder head design. They rotate at half engine speed and lack the inertia forces of reciprocating valve mechanisms. This allows for higher engines speeds offering appr. 10% more power. The 1980's MGN W12 F1 engine used rotary valves but never raced. Between 2002 and 2004 the Australian developer *Bishop Innovation* and Mercedes-Ilmor tested rotary valves for a F1 V10 engine.

Use in Chromatography

Rotary valves are used for loading samples on columns used for liquid or gas chromatography. The valves used in these methods are usually 6-port, 2-position rotary valves.

Chapter 2

Thermal Expansion Valve

A thermal expansion valve (often abbreviated as TEV, TXV or TX valve) is a component in refrigeration and air conditioning systems that controls the amount of refrigerant flow into the evaporator thereby controlling the superheating at the outlet of the evaporator. Thermal expansion valves are often referred to generically as "metring devices".

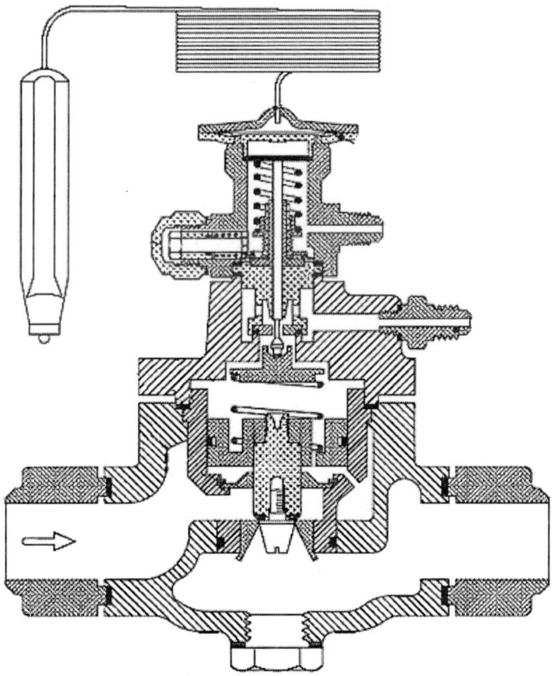

Figure: Thermostatic expansion valve

Flow control, or metring, of the refrigerant is accomplished by use of a temperature sensing bulb filled with a similar gas as in the system that causes the valve to open against the spring pressure in the valve body as the temperature on the bulb increases. As temperatures in the evaporator decrease, so does the pressure in the bulb and therefore on the spring causing the valve to close. An air conditioning system with a TX valve is often more efficient than other designs that do not use one.

A thermal expansion valve is a key element to a refrigeration cycle; the cycle that makes air conditioning, or air cooling, possible. A basic refrigeration cycle consists of four major elements, a compressor, a condenser, a metring device and an evaporator. As a refrigerant passes through a circuit containing these four elements, air conditioning occurs. The cycle starts when refrigerant enters the compressor in a low pressure, low temperature, gaseous form. The refrigerant is compressed by the compressor to a high pressure and temperature gaseous state. The high pressure and temperature gas then enters the condenser. The condenser precipitates the high pressure and temperature gas to a high temperature liquid by transferring heat to a lower temperature medium, usually ambient air. The high temperature liquid then enters the expansion valve where the TX valve allows a portion of the refrigerant to enter the evaporator. In order for the higher temperature fluid to cool, the flow must be limited into the evaporator to keep the pressure low and allow expansion back into the gas phase.

Pressure Regulator

A pressure regulator is a valve that automatically cuts off the flow of a liquid or gas at a certain pressure. Regulators are used to allow high-pressure fluid supply lines or tanks to be reduced to safe and/ or usable pressures for various applications.

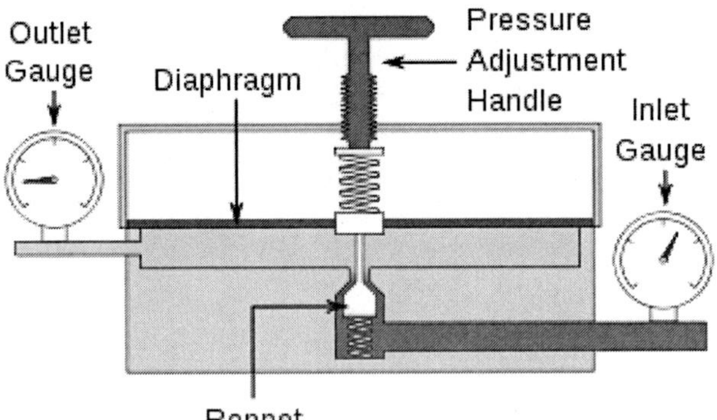

Figure: Single-stage pressure regulator

Gas pressure regulators are used to regulate the gas pressure and are not appropriate for measuring flow rates. Flowmetres, Rotometres or Mass Flow Controllers should be used to accurately regulate gas flow rates.

Operation

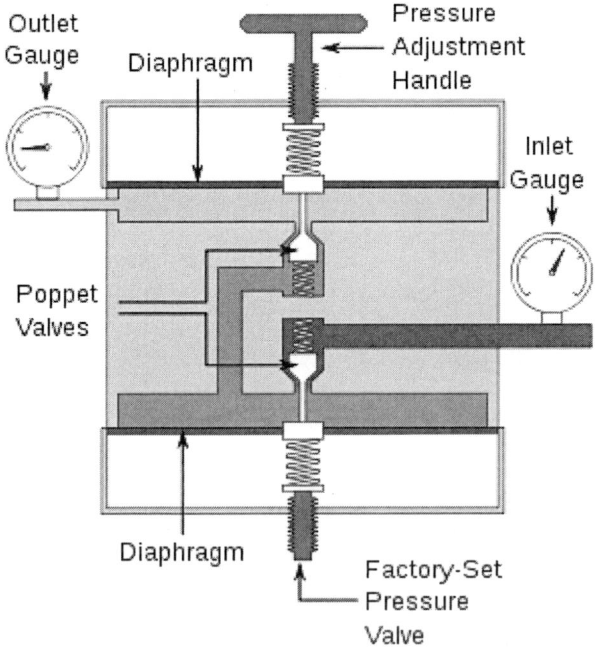

Figure: Two-stage pressure regulator

A pressure regulator's primary function is to match the flow of gas through the regulator to the demand for gas placed upon the system. If the load flow decreases, then the regulator flow must decrease also. If the load flow increases, then the regulator flow must increase in order to keep the controlled pressure from decreasing due to a shortage of gas in the pressure system.

A pressure regulator includes a *restricting element*, a *loading element*, and a *measuring element*:

- The restricting element is a type of valve. It can be a globe valve, butterfly valve, poppet valve, or any other type of valve that is capable of operating as a variable restriction to the flow.
- The loading element applies the needed force to the restricting element. It can be any number of things such as a weight, a spring, a piston actuator, or more commonly the diaphragm actuator in combination with a spring.
- When the actuator is forced against an expansion disk, the

force is distributed among the pressure walls. This allows the gas to flow at the proper rate and not to be continually vapourised and diluted.

- The measuring element determines when the inlet flow is equal to the outlet flow. The diaphragm is often used as a measuring element because it can also serve as a combine element.

In the pictured single-stage regulator, a diaphragm is used with a poppet valve to regulate pressure. As pressure in the upper chamber increases, the diaphragm is pushed upward, causing the poppet to reduce flow, bringing the pressure back down. By adjusting the top screw, the downward pressure on the diaphragm can be increased, requiring more pressure in the upper chamber to maintain equilibrium. In this way, the outlet pressure of the regulator is controlled.

Applications

Water Pressure Reduction

Often, water enters water-using appliances at fluctuating pressures, especially in remote locations, and industrial settings. This pressure often needs to be kept within a range to avoid damage to appliances, or accidents involving burst pipes/conduits. A single-stage regulator is sufficient in accuracy due to the high error tolerance of most such appliances.

Oxy-Fuel Welding and Cutting

Oxy-fuel welding and cutting processes require gases at specific pressures, and regulators will generally be used to reduce the high pressures of storage cylinders to those usable for cutting and welding. Oxy-gas regulators usually have two stages: The first stage of the regulator releases the gas at a constant rate from the cylinder despite the pressure in the cylinder becoming less as the gas is released. The second stage of the regulator controls the pressure reduction from the intermediate pressure to low pressure. It is constant flow. The valve assembly has two pressure gauges, one indicating cylinder pressure, the other indicating hose pressure.

Propane/LP Gas

All propane and LP Gas applications require the use of a regulator. Because pressures in propane tanks can fluctuate significantly, regulators must be present to deliver a steady flow pressure to downstream appliances. These regulators normally compensate for tank pressures between 30 - 200psig and commonly deliver 11 inches water column (0.4 psig) for residential applications and 35 inches of water column (1.3 psig) for industrial applications. Propane regulators differ in size and shape, delivery pressure and adjust-ability but are uniform in their purpose to deliver a constant outlet pressure for downstream requirements. As is the case in all regulators, outlet pressure is lower than inlet pressure.

Recreational Vehicles

For recreational vehicles with plumbing, a pressure regulator is a necessity. When camping, a source of water may have an enormous pressure level, particularly if it comes from a tank that is at a much higher elevation than the campground. Water pressure is dependent on how far the water must fall. Without a pressure regulator, the intense pressure encountered at some campgrounds in mountainous areas may be enough to burst the camper's water pipes or unseat the plumbing joints, causing flooding. Pressure regulators for this purpose are typically sold as small screw-on accessories that fit inline with the hoses used to connect an RV to the water supply, which are almost always screw-thread-compatible with the common garden hose.

Breathable Air Supply

Pressure regulators are used with air tanks for SCUBA diving. The tank may contain pressures well in excess of 2,000psig, which could cause a fatal barotrauma injury to a person breathing it directly. A regulator allows only a sustained flow of air at the ambient pressure (which varies by depth in the water).

Mining Industry

As the pressure builds rapidly in relation to depth, underground mining operations require a fairly complex water system with pressure reducing valves. These devices must be installed at a certain distance interval, usually 600 feet (180 m). Without such valves, pipes would easily burst and pressure would be too great for equipment operation.

Sampling Valve

A Sampling Valve is a type of valve used in process industries that allows taking a representative portion of a fluid (gases, liquids, fluidized, solids, or slurries) to test (e.g. by physical measurements, chemical analysis, microbiological examination), typically for the purposes of identification, quality control, or regulatory assessment. It is a valve used for sampling.

The Sampling Valve allows the operator to extract a sample of the product from the production line or reactor and safely store it for transportation to the laboratory where it will be analysed or to the archive room where it can be retrieved for further use.

In chemical plants, a sample can be taken during the process to ensure that the output meets specifications (or that the quality is acceptable), before shipping the chemical good or before accepting the chemical product.

Sampling Problem

When sampling chemical products, it is of utmost importance to select representative material to analyse. The sample must be representative of the lot, and the choice of samples is critical to producing a valid analysis. The statistics of the sampling process are also important.

And in order for a sample to be representative, it must not contain remains of previous batches that might have stay in the valve's "dead space". For example, a ball valve commonly found in tap valves consists of a ball that control the flow of the liquid. Once the valve is closed, liquid stays in the ball and the next time the valve is opened, that liquid flows first. If a sample is taken using a ball valve, the valve has to be flushed before the sample is taken.

This raises two additional problems:

1. If the liquid is a chemical substance that solidifies at room temperature such as polymer or other plastics, it can block the valve and prevent a second sample to be taken.
2. The residue of the chemical substance that was flushed needs to be recycled or destroyed at a cost both for the chemical plant and the environment.

When sampling hazardous chemicals or deadly substances, it is extremely important that the operator is not exposed to toxic fumes or vapours. These can fatally harm people and should they be released in the atmosphere, they will pollute the environment.

Choosing a Sample Valve

When choosing a Sampling Valve, different factors have to be considered:

- The material, pressure and temperature rating, gaskets, size and position of the valve according to the pipe specification of the plant.

- The handle or actuator type according to the position (reach) of the valve and depending of the chemical to sample (i.e. if the product contains solid particle or not)
- The receptacle to store the sample for transportation (bottles, container, bag, piston injector or syringe)

Safety Valve

A safety valve is a valve mechanism for the automatic release of a substance from a boiler, pressure vessel, or other system when the pressure or temperature exceeds preset limits.

It is part of a bigger set of pressure safety valves (PSV) or pressure relief valves (PRV). The other parts of the set are relief valves, safety relief valves, pilot-operated relief valves, low pressure safety valves, and vacuum pressure safety valves.

Safety valves were first used on steam boilers during the industrial revolution. Early boilers without them were prone to accidental explosion.

Vacuum safety valves (or combined pressure/vacuum safety valves) are used to prevent a tank from collapsing while emptying it or when cold rinse water is used after hot CIP or SIP. The calculation method is not defined in any norm when sizing a vacuum safety valve, particularly in the hot CIP / cold water scenario, but some manufacturers have developed simulations to do so.

Aspin Valve

An Aspin valve consists of a cone-shaped metal part fitted to the cylinder head. Aspin valves were first patented by Frank Metcalf Aspin in 1937 although the idea was probably devised before this time. It has a large opening in one side which reveals the gases in its internal combustion chamber to the spark plug, inlet port and outlet port accordingly. It is attached to the engine via a shaft at its top, allowing it to rotate horizontally above the cylinder.

Function and Design

The earliest and simplest safety valve on the steam digester in 1679 used a weight to hold the steam pressure (this design is still commonly used on pressure cookers); however, these were easily tampered with or accidentally released. On the Stockton and Darlington Railway, the safety valve tended to go off when the engine hit a bump in the track. A valve less sensitive to sudden accelerations used a spring to contain the steam pressure, but these (based on a Salter spring balance) could still be screwed down to increase the pressure beyond design limits. This dangerous practice was sometimes used to marginally increase the performance of a steam engine. In 1856 John Ramsbottom invented a tamper-proof spring safety valve that became universal on railways.

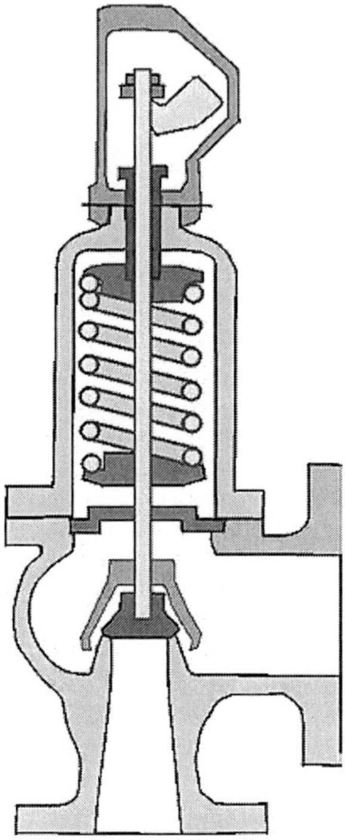

Figure: A cross-section of a proportional-safety valve

Safety valves also evolved to protect equipment such as pressure vessels (fired or not) and heat exchangers. The term safety valve should be limited to compressible fluid application (gas, vapour, or steam).

The two general types of protection encountered in industry are *thermal protection* and *flow protection.*

For liquid-packed vessels, thermal relief valves are generally characterised by the relatively small size of the valve necessary to provide protection from excess pressure caused by thermal expansion. In this case a small valve is adequate because most liquids are nearly incompressible, and so a relatively small amount of fluid discharged through the relief valve will produce a substantial reduction in pressure.

Flow protection is characterised by safety valves that are considerably larger than those mounted in thermal protection. They are generally sized for use in situations where significant quantities of gas or high volumes of liquid must be quickly discharged in order to protect the integrity of the vessel or pipeline. This protection can alternatively be achieved by installing a high integrity pressure protection system (HIPPS).

Technical Terms

In the petroleum refining, petrochemical, chemical manufacturing, natural gas processing, power generation, food, drinks, cosmetics and pharmaceuticals industries, the term safety valve is associated with the terms pressure relief valve (PRV), pressure safety valve (PSV) and relief valve.

The generic term is Pressure relief valve (PRV) or pressure safety valve (PSV) It should be noted that PRVs and PSVs are not the same thing, despite what many people think; the difference is that PSVs have a manual lever to open the valve in case of emergency.

- Relief valve (RV): automatic system that is actuated by static pressure in a liquid-filled vessel. It specifically opens proportionally with increasing pressure.
- Safety valve (SV): automatic system that relieves the static pressure on a gas. It usually opens completely, accompanied by a popping sound.
- Safety relief valve (SRV): automatic system that relieves by static pressure on both gas and liquid.
- Pilot-operated safety relief valve (POSRV): automatic system that relieves on remote command from a pilot to which the static pressure (from equipment to protect) is connected.
- Low pressure safety valve (LPSV): automatic system that relieves static pressure on a gas. Used when the difference between the vessel pressure and the ambient atmospheric pressure is small.
- Vacuum pressure safety valve (VPSV): automatic system that relieves static pressure on a gas. Used when the pressure difference between the vessel pressure and the ambient pressure is small, negative and near the atmospheric pressure.
- Low and vacuum pressure safety valve (LVPSV): automatic system that relieves static pressure on a gas. The pressure is small, negative or positive and near the atmospheric pressure.

RV, SV and SRV are spring-operated (even said spring-loaded). LPSV and VPSV are spring-operated or weight-loaded.

Legal and Code Requirements in Industry

In most countries, industries are legally required to protect pressure vessels and other equipment by using relief valves. Also, in most countries, equipment design codes such as those provided by the ASME, API and other organisations like ISO (ISO 4126) must be complied with and those codes include design standards for relief valves. Today, industries such as food, drinks, cosmetics, pharmaceuticals and fine chemicals industries ask for hygienic safety valves, fully drainable and Cleanable In Place. Most of these are stainless steel made; the hygienic norms are mainly 3A in the US and EHEDG in Europe.

Types

There is a wide range of safety valves having many different applications and performance criteria required to cover different areas. In addition, national standards are set by many kinds of safety valve.

- ASME I tap: A safety valve in accordance with the requirements of Section I of the application code ASME pressure vessel, which opens 3% and 4% of the pressure. Will rule on two rings serving and supported by a National Seal V defined.
- ASME VIII valve: Safety valve in accordance with the requirements of Article VIII of the ASME code for pressure vessel applications, which is within 10% overpressure that opens and closes in 7%. Characterised by a National Board UV stamp.
- Low-lift safety valve: The current position of the disc around the drain valve.
- Full lift safety valve: The region of the exemption is not determined by the position of the disc.
- Full-flow relief valve: A valve which is expected in the hole and lift the valve a sufficient measure of the minimum area for each position or under the seat to make the control panel.
- Safety Valve * classic: The spring housing is vented to the pressure side, i.e., functional characteristics are directly influenced by changes in pressure in the valve.
- Valve balanced: A balanced valve includes a means to minimise the effects of pressure on the operating characteristics of the valve.
- Guide pressure valve: The largest dump device is in conjunction with controlled, self-actuated auxiliary pressure relief.

- Power operated valve: A pressure relief valve in which the main system to relieve pressure combined with and controlled by a device requires an external power source.
- Is the standard safety valve: A valve opening to achieve the degree of buoyancy required for mass flow increases the pressure to reject over 10%. (The valve is characterised by a score of popular art, sometimes also called high-lift).
- Full-lift (solid line) valve: A valve that opens at the beginning of the elevator, quickly, with a 5% increase in pressure until complete removal is limited by design. The amount of lift to an early start (analog) can not exceed 20%.
- Valve and spring: A safety valve for the treatment of violent opening of the valve plate by a clamping force as a spring or weight.
- Proportional-relief valve: A safety valve which opens more or less stable in comparison with increasing pressure. Sudden opening within a stroke by 10% will not happen without increasing pressure. After opening at a pressure greater than 10% meet these safety valves of the lift and landed on the mass flowmetre.
- Safety valve diaphragm: A direct-loaded safety valve with linear and rotary components and springs to protect against the effects of the liquid membrane.
- Bellows valve: A direct-loaded safety valve and slide (or part) of rotating elements and sources are protected from the effects of fluid through bellows. The bellows allows such an interpretation, it is as if the influences of pressure.
- Controlled valve: A valve that consists of a main valve and a control unit. It also includes direct acting safety valves with additional load, which, until the total pressure reached, an additional force increases the closing force.
- Safety valve: A valve that automatically, without using any form of energy than the liquid in the discharge quantity of liquid, so as to prevent a predetermined safe pressure is higher, and closed again to prevent the flow of more volatile after the normal working pressure were restored. Note that the valve can be characterised by a pop-action (quick opening) or open relationship (not necessarily linear) on increasing pressure over the whole pressure.

- Valve and spring: A safety valve when the charger through the liquid under pressure valve plate with mechanical pressure directly, unlike, for example, a weight, lever and weight or spring. * Assisted valve - A valve which can be lifted by a support mechanism for moving, also gradually the pressure that the pressure setting, and for failure to comply with a mechanism to support all requirements for safety valves in the standard.
- Safety valve spring in addition: As a safety valve until the inlet pressure safety valve to the pressure achieved additional power increases the power of foreclosure. Note that this extra power (extra weight), which is made from an external power source is available, reliable published when the inlet pressure reaches the safety valve pressure set. The amount of the extra load is placed so that if such a charge is released, the safety valve of certified capacity at a pressure not greater than 1. Protected 1 times the maximum allowable pressure devices.
- Master valve: A valve whose operation is initiated and controlled by the fluid discharged from a pilot valve is a safety net direct costs, provided that standard.

United States

- ASME (American Society of Mechanical Engineers) Boiler & Pressure Vessel Code, Section I
- ASME (American Society of Mechanical Engineers) Boiler & Pressure Vessel Code, Section VIII, Division 1
- API (American Petroleum Institute) Recommended Practice 520 and API Standard 526, API Standard 2000 (low pressure - Storage tank)

European Union

- ISO 4126 (harmonized with European Union directives)
- EN 764-7 (former CEN standard, harmonized with European Union directives, replaced with EN ISO 4126-1)
- AD Merkblatt (German)
- PED 97/23/EC (Pressure Equipment Directive - European Union)

Japan

- JIS: Japanese Industrial Standards

South Korea

- KOSHA.

Water Heaters

They are required on water heaters, where they prevent disaster in certain configurations in the event that a thermostat should fail. There are still occasional, spectacular failures of older water heaters that lack this equipment. Houses can be levelled by the force of the blast.

Pressure Cookers

Pressure cookers are pots for cooking with a pressure-proof lid. Cooking at pressure allows the temperature to rise above the normal boiling point of water (100 degrees Celsius at sea level) which speeds up the cooking and makes the cooking more thorough.

Pressure cookers usually have two safety valves. One is a hole upon which a weight sits. The other is a sealed rubber grommet which is ejected in a controlled explosion if the first valve gets blocked.

Ballcock

A ballcock (also balltap or float valve) is a mechanism or machine for filling water tanks, such as those found in flush toilets, while avoiding overflow and (in the event of low water pressure) backflow. The modern ballcock was invented by Joseph Bramah and Thomas Twyford.. It consists of a valve connected to a hollow sealed float by means of a lever, mounted near the top of the tank. The valve is connected to the incoming water supply, and is opened and closed by the lever which has the float mounted on the end. When the water level rises, the float rises with it; once it rises to a pre-set level, the mechanism forces the lever to close the valve and shut off the water flow. This is an example of negative feedback and of proportional control.

When a toilet's handle is turned, a flapper valve is pulled upward by means of a rod or chain. It is designed to sink more slowly than the water, which will exit to the toilet bowl below, so that the tank may empty. As the tank water level drops, the float descends and actuates the fill valve. Water is fed to the tank to replenish its supply, and a smaller flow is directed into the overflow tube to refill the bowl.

Once the flapper valve closes, the water flow from the fill valve continues until the tank level again reaches the fill line. Should the float or valve fail and allow the water level to exceed the fill line, the water will pour into the overflow tube and out to the bowl. (Note that although this wastes water, it is a successful attempt to mitigate the flooding that would otherwise occur under that single-point valve failure.)

Tap (Valve)

A tap (also called spigot and faucet in the U.S. and Canada) is a valve controlling release of liquids or gas.

Water Taps

Figure: Water spigot; also known as a valve, hose hydrant, hose bibb, or sillcock.

Figure: Old metal taps were often part of decoration as this one in Fužine castle backyard, Slovenia.

Figure: Plastic taps are widely used in Kerala.

Water for baths, sinks and basins can be provided by separate hot and cold taps; this arrangement is common in older installations, particularly in public washrooms/lavatories and utility rooms/laundries. In kitchens and bathrooms mixer taps are commonly used. In this case, hot and cold water from the two valves is mixed before reaching the outlet, allowing the water to emerge at any temperature between that of the hot and cold water supplies. Mixer taps were invented by Thomas Campbell of Saint John, New Brunswick and patented in 1880.

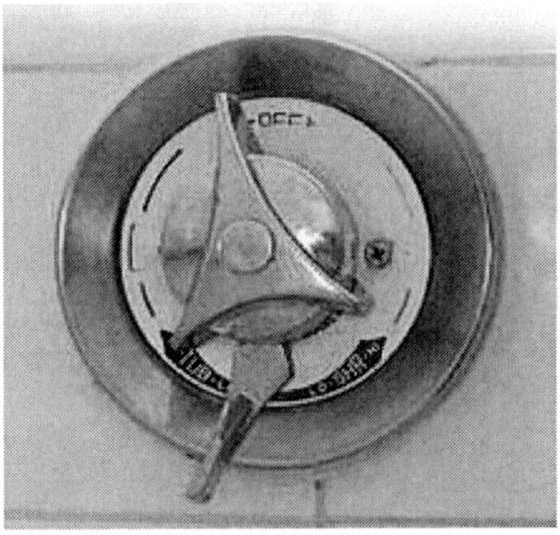

Figure: North American shower tap. Lower lever controls water exit; right: to shower ("SHR"), left: to bathtub ("TUB"), middle: no water. Middle lever temperature controls: turn counterclockwise to augment water flow, turn further to increase temperature.

For baths and showers, mixer taps frequently incorporate some sort of pressure balancing feature so that the hot/cold mixture ratio will not be affected by transient changes in the pressure of one or the other of the supplies. This helps avoid scalding or uncomfortable chilling as other water loads occur (such as the flushing of a toilet). Rather than two separate valves, mixer taps frequently use a single, more complex, valve controlled by a single handle (single handle mixer). The handle moves up and down to control the amount of water flow and from side to side to control the temperature of the water. Especially for baths and showers, the latest designs do this using a built in thermostat. These are known as thermostatic mixing valves, or TMVs, and can be mechanical or electronic. There are also faucets with colour LEDs to show the temperature of the water.

If separate taps are fitted, it may not be immediately clear which tap is hot and which is cold. The hot tap generally has a red indicator while the cold tap generally has a blue or green indicator. In the United States, the taps are frequently also labelled with an "H" or

"C". Note that in countries with Romance languages, the letters "C" for hot and "F" for cold are used (from French "chaud"/Italian "caldo" (hot) and French "froid"/Italian "freddo" (cold)). This can create confusion when English speakers visit these countries or vice versa. Mixer taps may have a red-blue stripe or arrows indicating which side will give hot and which cold.

In most countries, there is a standard arrangement of hot/cold taps. For example, in the United States and many other countries, the hot tap is on the left by building code requirements. Many installations exist where this standard has been ignored (called "crossed connections" by plumbers). Mis-assembly of some single-valve mixer taps will exchange hot and cold even if the fixture has been plumbed correctly. Most handles on residential homes are connected to the valve shaft and fastened down with a screw. Although on most commercial and industrial applications they are fitted with a removable key called a "loose key", "water key", or "sillcock key", which has a square peg and a square ended key to turn off and on the water. You can also take off the "loose key" to prevent vandals from turning on the water. In older buildings before the "loose key" was invented it was common for some landlords or caretakers to take off the handle of a residential tap, which had teeth that would meet up with the gears on the valve shaft. This Teeth and cog system is still used on most modern faucets. Although most of the time a "loose key" is on industrial and commercial applications sometimes you may see a "loose key" on homes by the seashore to prevent passers-by from washing the sand off their feet.

Beer taps: While in other contexts, depending on location, a "tap" may be a "faucet", "valve" or "spigot", the use of *"tap"* for beer is almost universal. This may be because the word was originally coined for the wooden valve in traditional barrels. Draft beer dispensed with a valve is said to be "on tap" (as an idiom). A "beer tap" now may be one of several items:

Pressure-Dispense Bar Tap: Almost universally in modern times, bulk beer is supplied in kegs that are served with the aid of external pressure. In a normal bar dispense system, this pressure comes from a cylinder of carbon dioxide in North America or nitrogen in the UK which forces the beer out of the keg and up a narrow tube to the bar. At the end of this tube is a valve built into a fixture (usually somewhat decorative) on the bar. This is the beer tap, and opening it with a small lever causes beer, pushed by the gas from the cylinder, to flow into the glass.

Portable Keg Tap: Sometimes, beer kegs designed to be connected to the above system are instead used on their own, perhaps at a party or outdoor event. In this case, a self-contained portable tap is required that allows beer to be served straight from the keg. Because the keg system uses pressure to force the beer up and out of the keg, these taps must have a means of supplying it. The typical "picnic tap" uses a hand pump to push air into the keg; this will cause the beer to spoil faster but is perfectly acceptable when it will be consumed in a short time. Portable taps with small CO_2 cylinders are also available.

Cask Beer Tap: Beers brewed and served in the traditional way (typically cask ale) do not use artificial gas. Taps for cask beer are simple on-off valves that are hammered into the end of the cask. When beer is served directly from the cask ("by gravity"), as at beer festivals and some pubs, it simply flows out of the tap and into the glass. When the cask is stored in the cellar and served from the bar, as in most pubs, the beer line is screwed onto the tap and the beer is sucked through it by a hand-operated low-pressure pump on the bar. The taps used are the same, and in beer-line setups the first pint is often poured from the cask as for "gravity", for tasting, before the line is connected. Cask beer taps can be brass (now discouraged for fear of lead contamination), stainless steel (good, but expensive), plastic (acceptable, and cheaper), and wood (to be avoided if possible).

Gas Taps

Although a gas tap may be a valve that releases any gas, the word is most commonly used to refer to taps that control the flow of fuel gas (natural gas or, historically, coal gas, syngas, etc.) in the home (for gas fires or other appliances) or in laboratories (for Bunsen burners).

Physics of Taps

Most water and gas taps have adjustable flow. Turning the knob or working the lever sets the flow rate by adjusting the size of an opening in the valve assembly, giving rise to choked flow through the narrow opening in the valve. The choked flow rate is independent of the viscosity or temperature of the fluid or gas in the pipe, and depends only weakly on the supply pressure, so that flow rate is stable at a given setting. At intermediate flow settings the pressure at the valve restriction drops nearly to zero from the venturi effect; in water taps, this causes the water to boil momentarily at room temperature as it passes through the restriction. Bubbles of cool water vapour form and collapse at the restriction, causing the familiar hissing sound. At very low flow settings, the viscosity of the water becomes important and the pressure drop (and hissing noise) vanish; at full flow settings, parasitic drag in the pipes becomes important and the water again becomes quiet.

One reason that most beer taps are not designed for adjustable flow is that the beer itself is damaged by the pressure drop in a choked-flow valve: holding a beer tap partially open causes the beer to foam vigorously, ruining the pour.

Tap Mechanisms

The first screw-down tap mechanism was patented and manufactured by the Rotherham brass founders Guest and Chrimes in 1845. Most older taps use a soft rubber or neoprene washer which is screwed down onto a valve seat in order to stop the flow. This is called a "globe valve" in engineering and, while it gives a leak-proof seal and good fine adjustment of flow, both the rubber washer and the valve seat are subject to wear (and for the seat, also corrosion) over time, so that eventually no tight seal is formed in the closed position, resulting in a leaking tap. The washer can be replaced and the valve seat resurfaced (at least a few times), but globe valves are never maintenance-free.

Also, the tortuous S-shaped path the water is forced to follow offers a significant obstruction to the flow. For high pressure domestic water systems this does not matter, but for low pressure systems where flow rate is important, such as a shower fed by a storage tank, a "stop tap" or, in engineering terms, a "gate valve" is preferred.

Gate valves use a metal disc the same diametre as the pipe which is screwed into place perpendicularly to the flow, cutting it off. There is no resistance to flow when the tap is fully open, but this type of tap rarely gives a perfect seal when closed. In the UK this type of tap normally has a wheel-shaped handle rather than a crutch or capstan handle.

Cone valves or ball valves are another alternative. These are commonly found as the service shut-off valves in more-expensive water systems and usually found in gas taps (and, incidentally, the cask beer taps referred to above). They can be identified by their range of motion—only 90°—between fully on and fully off. Usually, when the handle is in line with the pipe the valve is on, and when the handle is across the pipe it is closed. A cone valve consists of a shallowly-tapering cone in a tight-fitting socket placed across the flow of the fluid. In UK English this is usually known as a *taper-plug cock*. A ball valve uses a spherical ball instead. In either case, a hole through the cone or ball allows the fluid to pass if it is lined up with the openings in the socket through which the fluid enters and leaves; turning the cone using the handle rotates the passage away, presenting the fluid with the unbroken surface of the cone through which it cannot pass.

Valves of this type using a cylinder rather than a cone are sometimes encountered, but using a cone allows a tight fit to be made even with moderate manufacturing tolerances. The ball in ball valves rotates within plastic seats.

Hands free infrared proximity sensors are replacing the standard valve. Thermostatically controlled electronic dual-purpose mixing or diverting valves are used within industrial applications to automatically provide liquids as required.

Foot controlled valves are installed within laboratory and healthcare/hospitals. Modern taps often have aerators at the tip to help save water and reduce splashes. Without an aerator, water usually flows out of the tap in one big stream. An aerator spreads the water flow into many small droplets.

Modern bathroom and kitchen taps often use ceramic or plastic surfaces sliding against other spring-loaded ceramic surfaces or plastic washers. These tend to require far less maintenance than traditional globe valves and when maintenance is required, the entire interior of the valve is usually replaced, often as a single pre-assembled *cartridge.*

Of the trio of well-respected faucet manufacturers in North American plumbing circles, Moen and American Standard use cartridges (Moen's being O-ring based, American Standard's being ceramic), while Delta uses easily-replaced rubber seats facing the cartridge(s). Each design has its advantages: Moen cartridges tend to be easiest to find, American Standard cartridges have nearly infinite lifespan in sediment-free municipal water, and Delta's rubber seats tend to be most forgiving of sediment in well water.

Backflow

Most U.S. jurisdictions now require bibcocks to have a vacuum breaker or backflow preventer, so that water cannot return through the bibcock from the hose. This prevents contamination of the building or public water system should there be a pressure drop. In the UK, a double check valve is required to conform with water regulations; this is often incorporated within the body of the tap itself.

Nomenclature and Usage Variations

In the British Isles and most of the Commonwealth, the word "tap" is used for any everyday type of valve, particularly the fittings that control water supply to bathtubs and sinks. In the U.S., the word is more often used for beer taps, cut-in connections, or wiretapping. "Spigot" or "faucet" are more often used to refer to water valves, although this sense of "tap" is not uncommon, and the term "tap water" is the standard name for water from the faucet. Between "spigot" and "faucet", the connotative distinction is outdoor-versus-indoor, and utilitarian-versus-decorative; thus a spigot is an outdoor tap such as the bibcock (sillcock) for a garden hose, whereas a faucet is an indoor tap such as on the kitchen sink, bathroom sink, or bathtub.

Blast Valve

A blast valve is used to protect a shelter, such as a fallout shelter or bunker, from the effects of sudden outside air pressure changes. A nuclear weapon creates a shock wave, which may produce sudden pressure changes of more than an atmosphere (about 15 pounds per square inch) even several miles or kilometres from the detonation point. After the shock wave passes, a sudden negative pressure follows.

If such pressure waves enter a shelter, they will likely do substantial harm to occupants and equipment. A blast valve is placed in air intake/exhaust pipes, that remains open normally, but automatically closes when strong pressure is applied in either direction.

A typical blast valve has entries of 6-12" diametre (150–300 mm), and a larger centre section. Within the centre section is a disk mounted on an axle, with weak springs that keep it centred, away from both entries. Pressure displaces the disk along the axle, until it plugs one entry or the other. After the blast, the springs return the disk to the centre, re-opening the valve. Another type of blast valve employs rounded metal tubes held in place by springs. As pressure rises it pushes the tubes against the frame of the valve closing the opening. These valves are typically unidirectional in airflow and are employed in large wall structures where large amounts of airflow is required.

Stopcock

A stopcock is a valve used to restrict or isolate the flow of a liquid or gas through a pipe. In Great Britain a stopcock, not to be confused with a gate valve or a DiCiaccio branch, is used to prevent flow of water into a domestic water system. There are usually two stopcocks for a home. One is usually found just outside the property boundary and can be used to isolate the building from the water supply. The other is inside the property where the supply enters the property. These valves are provided to allow maintenance and prevent flooding if the domestic water system is pierced.

Laboratory Glassware

For laboratory glassware, items sometimes have plug valves with conically-tapered inserts called stopcocks (sometimes referred to as "plugs" by laypeople). The valve bodies are usually made of glass, while the plugs are made of glass or Teflon. When the plug is made of glass, the handle and plug are fused together in one piece out of glass. When both valve body and plug are made of glass, grease is usually used to give a good seal, as well as to prevent the valve from seizing and subsequently cracking. Special stopcocks are available, such as a double-oblique design used in Schlenk lines which permit the application of inert gas and vacuum from the same stopcock.

Double Beat Valve

The double-beat valve, drop valve or equilibrium valve is a type of poppet valve arranged to allow it to be opened against a high pressure with a minimum of force.

One of its uses is in steam engines to admit steam to the cylinders and to release the exhaust. In stationary steam engines it is usually operated by trip valve gear while in railway locomotives a valve gear such as Caprotti is used.

It was invented around 1800 by Jonathan Hornblower.

His valve is in the form of a hollow cylinder provided with two seats of nearly equal diametre, at *A* and *B* in the diagram. The cylinder is connected to the actuating rod by a web. The force required to lift the valve depends on the difference between diametres *D* and *d*: the smaller this difference can be made, the less the force necessary to open the valve.

In Hornblower's valve, the steam flows past seat *B* only. In later developments of the valve, the central portion is waisted so that the flow can pass at *A* as well, so increasing the valve opening area for the same amount of lift.

Double Check Valve

A double check valve or double check assembly (DCA) is a backflow prevention device designed to protect water supplies from contamination. It is also a valve used in air brake systems on heavy trucks.

Usage in Water Supply: In the case of usage in water supply, it consists of two check valves assembled in series. This employs two operating principles: firstly one check valve will still act, even if the other is jammed wide open. Secondly the closure of one valve reduces the pressure differential across the other, allowing a more reliable seal and avoiding even minor leakage.

Small valves may be so compact as to be barely noticeable, particularly when they are integrated into the bodies of existing taps (faucets). Larger check valves may be installed with ball valves at the ends, for isolation and testing. Often, test cocks (very small ball valves) are in place to attach test equipment for evaluating whether the double check assembly is still functional.

The double check valve assembly is suitable for prevention of back pressure and back siphonage, but is not suitable for high hazard applications. It is commonly used on lawn irrigation, fire sprinkler and combi-boiler systems. If the hazard is higher, even a relatively low hazard such as using antifreeze in the fire sprinkler system, then a more reliable check valve such as a Reduced pressure zone device may be mandated.

Flow Control Valve

A flow control valve regulates the flow or pressure of a fluid. Control valves normally respond to signals generated by independent devices such as flow metres or temperature gauges.

Control valves are normally fitted with actuators and positioners. Pneumatically-actuated globe valves and Diaphragm Valves are widely used for control purposes in many industries, although quarter-turn types such as (modified) ball, gate and butterfly valves are also used.

Control valves can also work with hydraulic actuators (also known as hydraulic pilots). These types of valves are also known as Automatic Control Valves. The hydraulic actuators will respond to changes of pressure or flow and will open/close the valve. Automatic Control Valves do not require an external power source, meaning that the fluid pressure is enough to open and close the valve. Automatic control valves include: pressure reducing valves, flow control valves, back-pressure sustaining valves, altitude valves, and relief valves. An altitude valve controls the level of a tank. The altitude valve will remain open while the tank is not full and it will close when the tanks reaches its maximum level. The opening and closing of the valve requires no external power source (electric, pneumatic, or man power), it is done automatically, hence its name.

Process plants consist of hundreds, or even thousands, of control loops all networked together to produce a product to be offered for sale. Each of these control loops is designed to keep some important process variable such as pressure, flow, level, temperature, etc. within a required operating range to ensure the quality of the end product. Each of these loops receives and internally creates disturbances that detrimentally affect the process variable, and interaction from other loops in the network provides disturbances that influence the process variable.

To reduce the effect of these load disturbances, sensors and transmitters collect information about the process variable and its relationship to some desired set point. A controller then processes this information and decides what must be done to get the process variable back to where it should be after a load disturbance occurs. When all the measuring, comparing, and calculating are done, some type of final control element must implement the strategy selected by the controller. The most common final control element in the process control industries is the control valve. The control valve manipulates a flowing fluid, such as gas, steam, water, or chemical compounds, to compensate for the load disturbance and keep the regulated process variable as close as possible to the desired set point.

Flutter Valve

A flutter valve (also known as the Heimlich valve after its inventor, Henry Heimlich) is a one-way valve used in respiratory medicine to prevent air from travelling back along a chest tube. It is most commonly used to help remove air from a pneumothorax. The valve is usually designed as a rubber sleeve within a plastic case where the rubber sleeve is arranged so that when air passes through the valve one way the sleeve opens and lets the air through. However when air is sucked back the other way the sleeve closes off and no air is allowed backwards. This construction enables it to act as a one-way valve allowing air (or fluid) to flow only one way along the drainage tube. The end of the drainage tube is placed inside the patient's chest cavity, within the air or fluid to be drained. The flutter valve is placed in the appropriate orientation (most packages are designed so the valve can only be connected in the appropriate orientation) and the pneumothorax is thus evacuated from the patient's chest.

There are several potential problems with these valves. One is that the chest tube can clog. When chest tube clogging occurs, the pneumothorax or subcutaneous emphysema can recur. This can also lead to empyema. The other is that these tend to leak fluid. To address this, some have turned to small chest drainage alternatives, such as the Atrium Mini or the Teleflex Mini Sahara. An alternative solution is to attach a sputum trap to the valve, thus providing a reservoir to capture the draining fluid.

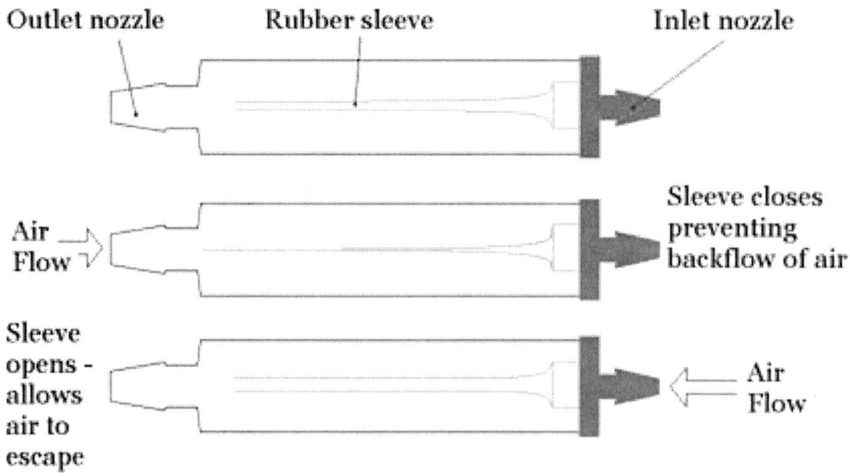

Figure: Flutter valve allowing air to flow one way

Heart Valve

A heart valve normally allows blood flow in only one direction through the heart. The four valves commonly represented in a mammalian heart determine the pathway of blood flow through the heart. A heart valve opens or closes incumbent upon differential pressure on each side.

The four valves in the heart are:

- The two atrioventricular (AV) valves, which are between the atria and the ventricles, are the mitral valve and the tricuspid valve.
- The two semilunar (SL) valves, which are in the arteries leaving the heart, are the aortic valve and the pulmonary valve.

A form of heart disease occurs when a valve malfunctions and allows some blood to flow in the wrong direction. This is called regurgitation.

Atrioventricular or Cuspid Valves

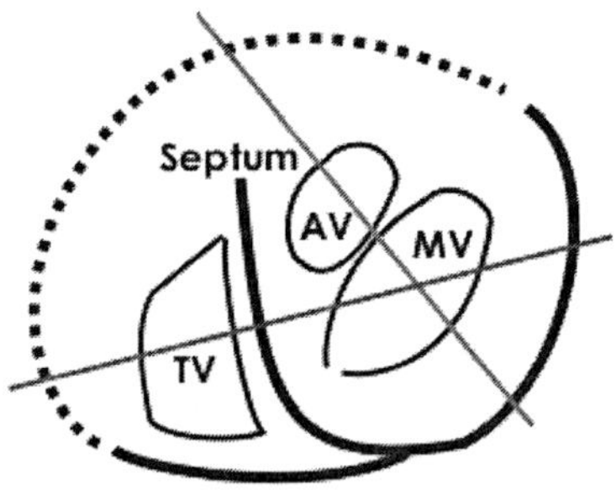

This is an explanation of the echocardiogram above. MV: Mitral valve, TV: Tricuspid valve, AV: Aortic valve, Septum: Interventricular septum. Continuous lines demarcate septum and free wall seen in echocardiogram, dotted line is a suggestion of where the free wall of the right ventricle should be. The red line represents where the upper left loop in the echocardiogram transects the 3D-loop, the blue line represents the lower loop.

These are small valves that prevent backflow from the ventricles into the atrium during systole. They are anchored to the wall of the ventricle by chordae tendineae, which prevent the valve from inverting.

The chordae tendineae are attached to papillary muscles that cause tension to better hold the valve. Together, the papillary muscles and the chordae tendineae are known as the subvalvular apparatus. The function of the subvalvular apparatus is to keep the valves from prolapsing into the atria when they close. The subvalvular apparatus have no effect on the opening and closure of the valves, however. This is caused entirely by the pressure gradient across the valve. The peculiar insertion of chords on the leaflet free margin however provides systolic stress sharing between chords according to their different thickness. The closure of the AV valves is heard as the first heart sound (S1). A heart valve normally allows blood flow in only one direction through the heart. The four valves commonly represented in a mammalian heart determine the pathway of blood flow through the heart. A heart valve opens or closes incumbent upon differential pressure on each side.

The four valves in the heart are:

- The two atrioventricular (AV) valves, which are between the atria and the ventricles, are the mitral valve and the tricuspid valve.

- The two semilunar (SL) valves, which are in the arteries leaving the heart, are the aortic valve and the pulmonary valve.

A form of heart disease occurs when a valve malfunctions and allows some blood to flow in the wrong direction. This is called regurgitation.

An artificial heart valve may be used to surgically replace a patient's damaged valve.

These are small valves that prevent backflow from the ventricles into the atrium during systole. They are anchored to the wall of the ventricle by chordae tendineae, which prevent the valve from inverting.

The chordae tendineae are attached to papillary muscles that cause tension to better hold the valve. Together, the papillary muscles and the chordae tendineae are known as the subvalvular apparatus. The function of the subvalvular apparatus is to keep the valves from prolapsing into the atria when they close. The subvalvular apparatus have no effect on the opening and closure of the valves, however. This is caused entirely by the pressure gradient across the valve. The peculiar insertion of chords on the leaflet free margin however provides systolic stress sharing between chords according to their different thickness. The closure of the AV valves is heard as the first heart sound (S1).

The tricuspid valve, or right atrioventricular valve, is on the right dorsal side of the mammalian heart, between the right atrium and the right ventricle. The normal tricuspid valve usually has three leaflets and three papillary muscles. They are connected to the papillary muscles by the chordae tendineae, which lie in the right ventricle. Tricuspid valves will not always consist of three leaflets and may also occur with two or four leaflets; the number may change during one's lifetime.

Diseases

Tricuspid regurgitation is not uncommon in the tricuspid valve.

It is a common valve to be infected (endocarditis) in IV drug users. Patients who inject narcotics or other drugs intravenously may introduce infection, which will travel to the right side of the heart, most often caused by S. aureus. In other patients without a history of intravenous exposure, endocarditis is more frequently left-sided.

The tricuspid valve can be affected by rheumatic fever, which can cause tricuspid stenosis or tricuspid insufficiency (also called tricuspid regurgitation). Some patients are born with congenital abnormalities of the tricuspid valve. Congenital apical displacement of the tricuspid valve is called Ebstein's anomaly and typically causes significant tricuspid regurgitation.

Aortic Valve

The aortic valve is one of the valves of the heart. It is normally tricuspid (with three leaflets), although in 1% of the population it is found to be congenitally bicuspid (two leaflets). It lies between the left ventricle and the aorta.

Function and Physiology

During ventricular systole, pressure rises in the left ventricle. When the pressure in the left ventricle rises above the pressure in the aorta, the aortic valve opens, allowing blood to exit the left ventricle into the aorta. When ventricular systole ends, pressure in the left ventricle rapidly drops. When the pressure in the left ventricle decreases, the aortic pressure forces the aortic valve to close. The closure of the aortic valve contributes the A_2 component of the second heart sound (S_2).

Disease of the Aortic Valve

There are two protypical processes that can affect the aortic valve - aortic stenosis in which the valve fails to open fully, thereby obstructing blood flow out from the heart, and aortic insufficiency, also called aortic regurgitation, in which the aortic valve is incompetent and blood flows passively back to the heart in the wrong direction. These two conditions frequently co-exist.

- Common causes of aortic stenosis include rheumatic fever, degenerative calcification, and congenital diseases such as bicuspid aortic valve.
- Common causes of aortic regurgitation include dilation of the aorta, previous rheumatic fever, infection, i.e. infective endocarditis, myxomatous degeneration of the aortic valve, and Marfan's syndrome.

Bicuspid Aortic Valve

The most common congenital abnormality of the heart is the bicuspid aortic valve. In this condition, instead of three cusps, the aortic valve has two cusps. This condition is often undiagnosed until later in life when the person develops symptomatic aortic stenosis. Aortic stenosis occurs in this condition usually in patients in their 40s or 50s, an average of 10 years earlier than can occur in people with congenitally normal aortic valves.

Patients with Turner's Syndrome usually have associated bicuspid aortic valves.

Aortic Valve Replacement

Aortic valve replacement means that a patient's aortic valve is replaced by a different valve. The aortic valve can be affected by a range of diseases and require aortic valve replacement. The valve can become either leaky (regurgitant or insufficient) or stuck partially shut (stenotic). Aortic valve replacement currently requires open heart surgery. Research is being done now to develop valves that can be implanted using a catheter without open heart surgery. There are two basic types of artificial heart valve, mechanical valves and tissue valves. Tissue heart valves are usually made from animal tissues, either animal heart valve tissue or animal pericardial tissue. The tissue is treated to prevent rejection and to prevent calcification.

There are alternatives to animal tissue valves. In some cases, a human aortic valve can be implanted. These are called homografts. Homograft valves are donated by patients and recovered after the patient expires. The durability of homograft valves is probably the same as for porcine tissue valves. Another procedure for aortic valve replacement is the Ross procedure (after Donald Ross) or pulmonary autograft. The Ross procedure involves going to surgery to have the aortic valve removed and replacing it with the patient's own pulmonary valve. A pulmonary homograft (a pulmonary valve taken from a cadaver) or a valvular prothesis is then used to replace the patient's own pulmonary valve.

The first minimally invasive aortic valve surgery took place at the Cleveland Clinic in 1996.

Bicuspid Aortic Valve

A bicuspid aortic valve (BAV) is most commonly a congenital condition of the aortic valve where two of the aortic valvular leaflets fuse during development resulting in a valve that is bicuspid instead of the normal tricuspid configuration. Normally the only cardiac valve that is bicuspid is the mitral valve (bicuspid valve) which is situated between the left atrium and left ventricle. Cardiac valves play a crucial role in ensuring the unidirectional flow of blood from the atrium to the ventricles, or the ventricle to the aorta or pulmonary trunk.

Incidence/Prevalence

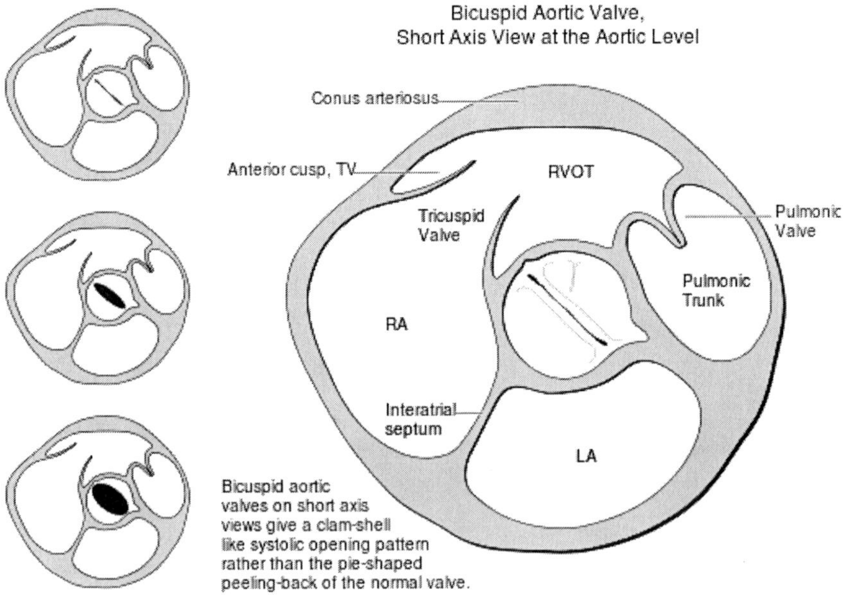

Figure: Heart bicuspid aortic valve anatomy.

Bicuspid aortic valves are the most common cardiac valvular anomaly, occurring in 1- 2% of the general population. It is twice as common in males as in females.

Bicuspid aortic valve is an inheritable condition, with a demonstrated association with Notch 1. Both familial clustering and isolated valve defects have been documented. The incidence of bicuspid aortic valve can be as high as 10% in families affected with the valve problem. Recent studies suggest that BAV is an autosomal dominant condition with incomplete penetrance. Other congenital heart defects are associated with bicuspid aortic valve at various frequencies, including coarctation of the aorta.

Presentation

In many cases, a bicuspid aortic valve will cause no problems. However BAV may become calcified later in life, which may lead to varying degrees of severity of aortic stenosis that will manifest as murmurs. If the leaflets do not close correctly, aortic regurgitation can occur. If these become severe enough, they may require heart surgery. People with BAV may become tired more easily than those with normal valvular function and have difficulty maintaining stamina for cardio-intensive activities-due to poor heart performance. The heart is put under more stress in order to either pump more blood through a stenotic valve or attempt to circulate regurgitation blood through a leaking valve.

Diagnosis, Treatment, and Prognosis

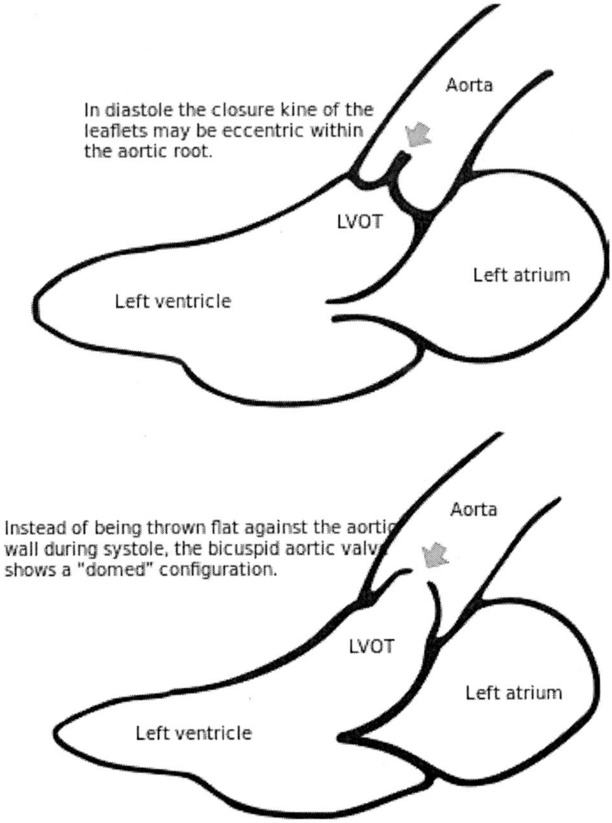

Figure: Heart bicuspid aortic valve diagram

The condition can be associated with a heart murmur located at the right second intercostal space. Often there will be differences in blood pressures between upper and lower extremities. The diagnosis can be assisted with echocardiography (EchoCG) or magnetic resonance imaging (MRI).

Most patients with bicuspid aortic valve whose valve becomes dysfunctional will need careful follow-up and potentially valve replacement at some point in life. Regular EchoCG and MRI may be performed. For diagnosed patients, genetic testing is done to allow for future offspring with the disease to be monitored and treated early in life.

Patients with bicuspid aortic valve should be followed by cardiologist or cardiac surgeon with specific interest in this valve pathology.

Average lifespan is similar to that of those without the anomaly.

Aorta Complications

One of the most notable associations with BAV is the tendency for these patients to present with ascending aortic aneurysmal lesions. The extracellular matrix of the aorta in patients with BAV shows marked deviations from that of the normal tricuspid aortic valve. It is currently believed that an increase in the ratio of MMP2 (Matrix Metalloproteinases 2) to TIMP1 (Tissue Inhibitor Metalloproteinases 1) may be responsible for the abnormal degradation of the valve matrix and therefore lead to aortic dissection and aneurysm. However, other studies have also shown MMP9 involvement with no differences in TIMP expression.

The size of the proximal aorta should be evaluated carefully during the work-up. The initial diametre of the aorta should be noted and annual evaluation with CT scan, or MRI to avoid ionizing radiation, should be recommended to the patient; the examination should be conducted more frequently if a change in aortic diametre is seen. From this monitoring, the type of surgery that should be offered to the patient can be determined based on the change in size of the aorta.

Co-arctation of the aorta (a congenital narrowing in the region of the ductus arteriosus) has also been associated with BAV.

Swirl Valve

Swirl valve technology has been developed by Twister BV of the Netherlands primarily for the gas processing market. This technology is similar to the swirl tube and can also be applied for liquid/liquid separation applications such as oil/water separation and for oil degassing. It is currently being used for optimising existing Joule–Thomson (JT-LTS) systems to minimise liquid carryover. The swirl valve is exactly the same as a Joule-Thomson (JT) choke valve, but it enhances the performance of downstream separators for the same pressure drop, by maximising droplet coalescence. The technology can be applied where a low temperature separator is undersized, or when a lower pressure drop over a JT valve is needed with a similar dew pointing performance. It can also be applied for reducing glycol/chemical inhibition liquid mist carry-over problems.

Artificial Heart Valve

N artificial heart valve is a device implanted in the heart of a patient with heart valvular disease. When one of the four heart valves malfunctions, the medical choice may be to replace the natural valve with an artificial valve. This requires open-heart surgery.

Valves are integral to the normal physiological functioning of the human heart. Natural heart valves are evolved to forms that perform the functional requirement of inducing unidirectional blood flow through the valve structure from one chamber of the heart to another. Natural heart valves become dysfunctional for a variety of pathological causes. Some pathologies may require complete surgical replacement of the natural heart valve with a heart valve prosthesis.

Types of Heart Valve Prostheses

There are two main types of artificial heart valves: the mechanical *and the* biological valves.

- Mechanical valves
  - o Percutaneous implantation
    - § Stent framed
    - § Not framed
  - o Sternotomy/Thoracotomy implantation
    - § Ball and cage
    - § Tilting disk
    - § Bi-leaflet
    - § Tri-leaflet
- Biological heart valves
  - o Allograft/isograft
  - o Xenograft

Mechanical Valves

Mechanical heart valves (MHV) are prosthetics designed to replicate the function of the natural valves of the human heart. The human heart contains four valves: tricuspid valve, pulmonic valve, mitral valve and aortic valve. Their main purpose is to maintain unimpeded forward flow through the heart and from the heart into the major blood vessels connected to the heart, the pulmonary artery and the aorta. As a result of a number of disease processes, both acquired and congenital, any one of the four heart valves may malfunction and result in either stenosis (impeded forward flow) and/

or backward flow (regurgitation). Either process burdens the heart and may lead to serious problems including heart failure. A mechanical heart valve is intended to replace a diseased heart valve with its prosthetic equivalent.

There are two basic types of valves that can be used for valve replacement, mechanical and tissue valves. Modern mechanical valves can last indefinitely (the equivalent of over 50,000 years in an accelerated valve wear tester). However, current mechanical heart valves all require lifelong treatment with anticoagulants (blood thinners), e.g. warfarin, which requires monthly blood tests to monitor. This process of thinning the blood is called anticoagulation. Tissue heart valves, in contrast, do not require the use of anticoagulant drugs due to the improved blood flow dynamics resulting in less red cell damage and hence less clot formation. Their main weakness however, is their limited lifespan. Traditional tissue valves, made of pig heart valves, will last on average 15 years before they require replacement (but typically less in younger patients).

Types of Mechanical Heart Valves

There are three major types of mechanical valves - *caged-ball*, *tilting-disk* and *bileaflet* - with many modifications on these designs.

The first artificial heart valve was the caged-ball, which utilises a metal cage to house a silicone elastomer ball. When blood pressure in the chamber of the heart exceeds that of the pressure on the outside of the chamber the ball is pushed against the cage and allows blood to flow. At the completion of the heart's contraction, the pressure inside the chamber drops and is lower than beyond the valve, so the ball moves back against the base of the valve forming a seal. In 1952, Dr. Charles A. Hufnagel implanted caged-ball heart valves in ten patients (six survived the operation), marking the first long-term success in prosthetic heart valves. A similar valve was invented by Miles "Lowell" Edwards and Albert Starr in 1960 (commonly referred to as the Starr-Edwards Silastic Ball Valve). The first human implant was on Sept 21, 1960. It consisted of a silicone ball enclosed in a cage formed by wires originating from the valve housing. Caged ball valves have a high tendency to forming blood clots, so the patient must have a high degree of anti-coagulation, usually with a target INR of 2.5-3.5. Edwards Lifesciences discontinued production of the Starr-Edwards valve in 2007.

Soon after came tilting-disc valves. The first clinically available tilting disk valve was the Bjork-Shiley valve and has undergone several significant design changes since its introduction in 1969. Tilting disk valves have a single circular occluder controlled by a metal strut. They are made of a metal ring covered by a ePTFE fabric, into which the suture threads are stitched in order to hold the valve in place. The metal ring holds, by means of two metal supports, a disc which opens and closes as the heart pumps blood through the valve. The disc is usually made of an extremely hard carbon material (pyrolytic carbon), in order to allow the valve to function for years without wearing out. The Medtronic-Hall model is the most common tilting-disc design in the US. In some models of mechanical valves, the disc is divided into two parts, which open and close as a door..

St. Jude Medical is the leader in bileaflet valves, which consist of two semicircular leaflets that rotate about struts attached to the valve housing. This design was introduced in 1979 and while they take care of some of the issues that were seen in the other models, bileaflets are vulnerable to backflow and so they cannot be considered as ideal. Bileaflet valves do, however, provide much more natural blood flow than caged-ball or tilting-disc implants. One of the main advantages of these valves is that they are well tolerated by the body. Only a small amount of blood thinner is needed to be taken by the patient each day in order to prevent clotting of the blood when flowing through the valve.

These *bileaflet* valves have the advantage that they have a greater effective opening area (2.4-3.2 square cm c.f. 1.5-2.1 for the single-leaflet valves). Also, they are the least thrombogenic of the artificial valves.

Mechanical heart valves are today very reliable and allow the patient to live a normal life. Most mechanical valves last for at least 20 to 30 years..

Durability

Mechanical heart valves have been traditionally considered to be more durable in comparison to their bioprosthetic counterparts. The struts and occluders are made out of either pyrolytic carbon or titanium coated with pyrolytic carbon, and the sewing ring cuff is Teflon, polyester or dacron. The major load arises from transvalvular pressure generated at and after valve closure, and in cases where structural failure does happen, it is usually as a result of occluder impact on the components.

Impact wear and friction wear dictate the loss of material in MHV. Impact wear usually occurs in the hinge regions of bileaflets, between the occluder and ring in tilting-discs, and between the ball and cage in caged-ball valves. Friction wear occurs between the occluder and strut in tilting-discs, and between the leaflet pivots and hinge cavities in bileaflets.

MHV made out of metal are also susceptible to fatigue failure owing to the polycrystalline characteristic of metals, but this is not an issue with pyrolytic carbon MHV because this material is not crystalline in nature.

Cavitation

Cavitation is an event that can lead to MHV failure. While this has been a relatively rare occurrence, in 1988 the Edwards-Duramedics bileaflet had 46 reported failures in 20,000 implants related to cavitation damage. Since manufacturers have made cavitation testing an essential part of the design verification process. Cavitation is the rapid formation of vapourous microbubbles in the fluid due to a local drop of pressure below the vapourisation pressure at a given temperature. When conditions for cavitation are present bubbles will form and at the time of pressure recovery they will collapse or implode. This event will cause pressure or thermal shockwaves and fluid microjets which can damage a surface. These thermodynamic conditions are known to be the cause of MHV related erosion.

The valvular event that causes such cavitating conditions to exist is the closing mechanics of the MHV. Several causes of cavitation relating to valve closure have been identified. Squeeze flow is the term used to describe cavitation that is said to occur as the occluder approaches the housing during closure and fluid is squeezed between the occluder and the valve housing causing a low pressure formation. Water hammer is a term used to describe cavitation caused by the sudden stop of the valve occluder as it contacts the valve housing. This sudden retard of the fluid retrograde inertia is said to put the fluid under tension causing cavitation. Squeeze flow is said to form a cloud of bubbles at the circumferential lip of the occluder whereas water hammer is said to be seen as transient bubbles at the occlude housing.

For either event, cavitation occurs on the upstream side of valve. Clinically, cavitation is of primary concern in the mitral position. This position is especially harsh due to the sudden ventricular pressure rise which drives the valve closure against a low left atrial pressure which is said to be the worst case condition thus position for cavitation to occur. Cavitation is also suspected as a contributing factor in blood cell damage and increased risk of thromboembloic complications.

The temporal rate of change of the left ventricular, measured as a slope of the ventricular pressure curve (dP/dt) is regarded as the best indicator for cavitation potential. Most MHV investigated generate cavitation only when the dP/dt is well above the physiological range. However investigations have found that several tilting disc valves and only one bileaflet valve, the Edwards-Duromedics, generate cavitation within the physiological range. Investigations have repeatedly demonstrated that bileaflet valves, with the exception of the Edwards Duramedics design, cavitate only at dP/dt levels well above the physiological range.

Fluid Mechanics

Many of the complications associated with MHV can be explained through fluid mechanics. For example, thrombus formation is a debilitating side effect of high shear stresses created by the design of the valves. An ideal heart valve from an engineering perspective would produce minimal pressure drops, have small regurgitation volumes, minimise turbulence, reduce prevalence of high stresses, and not create flow separations in the vicinity of the valve. One measure of the quality of a valve is the effective orifice area (EOA), which can be calculated as follows:

$$EOA(\text{cm}^2) = \frac{Q_{rms}}{51.6\sqrt{\Delta p}}$$

where Q_{rms} is the root mean square systolic/diastolic flow rate (cm^3/s) and Δp is the mean systolic/diastolic pressure drop (mmHg). This is a measure of how much the prosthesis impedes blood flow through the valve. A higher EOA corresponds to a smaller energy loss. The performance index (PI) normalises the EOA by valve size and is a size-independent measure of the valve's resistance characteristics. Bileaflet valves typically have higher PI's than tilted-disc models, which in turn have higher PI's than caged-ball models. As blood flows through a prosthetic heart valve, a sudden pressure drop occurs across the valve due to the reduction in cross-sectional area within the valve housing. This can be quantified through the continuity equation and Bernoulli's equation:

$$A_1V_1 = A_2V_2$$

$$P_1 + \frac{1}{2}\rho_1 V_1^2 = P_2 + \frac{1}{2}\rho_2 V_2^2$$

where A represents the cross-sectional area, P is pressure, ρ is density, and V is the velocity. As cross-sectional area decreases in the valve, velocity increases and pressure drops as a result. This effect is more dramatic in caged-ball valves than in tilting-disc and bileaflet valves. A larger systolic pressure is required to drive flow forward in order to compensate for a large pressure drop, so it should be minimised.

Regurgitation is the sum of retrograde flow during the closing motion of the valve and leakage flow after closure. It is directly proportional to valve size and is also dependent on valve type. Typically, caged-ball valves have a low amount of regurgitation as there is very little leakage. Tilting-disc and bileaflet valves are comparable, with the bileaflet valves have a slightly larger regurgitation volume. Bioprosthetics prevail over MHV in this case, as they have virtually no regurgitation volume.

Turbulence and high shear stresses are also major issues with MHV, as they can fracture the valve housing or components, or induce blood damage. A large flow gradient can lead to these factors, so flow separation and stagnation should be as small as possible. High stresses are created at the edges of the annular jet in caged-ball valves, in narrow regions at the edges of the major orifice jet in tilting-disc valves, and in regions immediately distal to the valve leaflets in bileaflet valves. The implications of blood damage from these stresses are discussed in the next section.

The cavitation phenomenon can also be described using fluid mechanics. This can result from pressure oscillations, flow deceleration, tip vortices, streamline contraction, and squeeze jets [4]. This last cause is the most contributive factor to cavitation. The squeeze jets are formed when the valve is closing and the blood between the occluder and valve housing is "squeezed" out to create a high-speed jet. This in turn creates intense vortices with very low pressures that can lead to cavitation.

Blood Damage

One of the major drawbacks of mechanical heart valves is that patients with these implants require consistent anti-coagulation therapy. Clots formed by red blood cell (RBC) and platelet damage can block up blood vessels and lead to very serious consequences. Clotting occurs in one of three basic pathways: tissue factor exposure, platelet activation, or contact activation by foreign materials, and in three steps: initiation, amplification, and propagation.

In the tissue factor exposure path, initiation begins when cells are ruptured and expose tissue factor (TF). Plasma Factor (f) VII binds to TF and sets off a chain reaction which activates fXa and fVa which bind to each other to produce thrombin which in turn activates platelets and fVIII. The platelets activate by binding to the damaged tissue in the initiation phase, and fibrin stabilises the clot during the propagation phase.

The platelet activation pathway is triggered when stresses reach a level above 6 to 8 Pa (60–80 dyn/cm^2). The steps involved with this are less clearly understood, but initiation begins with the binding of vWF from the plasma to GPIb on the platelet. This is followed by a large influx of Ca ions, which activates the platelets. GPIIb-IIIa facilitates platelet-platelet adhesion during amplification. The propagation step is still under study.

Contact activation begins when fXII binds to a procoagulant surface. This in turn activates prekallikrein (PK) and high-molecular-weight kininogen (HK). Eventually, HKa-PK and HKa-fXI complexes form on the surface. In amplification, Hka-FXIa complexes activate fIX to fIXa, which in turn forms thrombin and platelets. Proteins buildup on the surface and facilitate platelet adhesion and tissue growth in the propagation stage.

All MHV models are vulnerable to thrombus formation due to high shear stress, stagnation, and flow separation. The caged-ball designs experience high stresses at the walls that can damage cells, as well as flow separation due to high-velocity reverse flow surrounded by stagnant flow. Tilting-disc valves have flow separation behind the valve struts and disc as a result of a combination of high velocity and stagnant flows. The bileaflet models have high stresses during forward and leakage flows as well as adjacent stagnant flow in the hinge area. As it turns out, the hinge area is the most critical part of bileaflets and is where the thrombus formation usually prevails.

In general, blood damage affects valves in both the mitral and aortic positions. High stresses during leakage flow in aortal valves result from higher transvalvular pressures, and high stresses occur during forward flow for mitral valves. Valvular thrombosis is most common in mitral prosthetics. The caged-ball model is better than the other two models in terms of controlling this problem, because it is at a lower risk for thrombosis and it is gradual when it does happen. The bileaflet is more adaptable to this problem than the tilting-disc model because if one leaflet stops working, the other can still function. However, if the hinge is blocked, both leaflets will stop functioning.

Because all models experience high stresses, patients with mechanical heart valve implants require anti-coagulation therapy.

Bioprosthetics are less prone to develop blood clotting, but the trade-off concerning durability generally favours their use in patients older than age 55.

Mechanical heart valves can also cause hemolytic anemia with hemolysis of the red blood cells as they pass through the valve.

Biological Valves

Biological valves are valves of animals, like pigs, which undergo several chemical procedures in order to make them suitable for implantation in the human heart. The porcine (or pig) heart is most similar to the human heart, and therefore represents the best anatomical fit for replacement. Implantation of a porcine valve is a type of xenotransplantation, also known as a xenograft, which means a transplant from one species (in this case a pig) to another. There are some risks associated with a Xenograft such as the human body's tendency to reject foreign material. Medication can be used to retard this effect, but is not always successful.

Another type of biological valve utilises biological tissue to make leaflets that are sewn into a metal frame. This tissue is typically harvested from the Pericardial Sac of either Bovine (cows) or Equine (horses). The pericardial sac is particularly well suited for a valve leaflet due to its extremely durable physical properties. This type of biological valve is extremely effective means of valve replacement. The tissue is sterilized so that the biological markers are removed, eliminating a response from the host's immune system. The leaflets are flexible and durable and do not require the patient to take blood thinners for the rest of their life.

The most used heart valves in the US and EU are those utilising tissue leaflets. Mechanical valves are more commonly used in Asia and Latin America. The following companies manufacture tissue heart valves: Edwards Lifesciences, Medtronic, St. Jude Medical, Sorin, ATS, 3F, CryoLife, and LifeNet Health.

Functional Requirements of Heart Valve Prostheses

The functioning of natural heart valves is characterised by many advantages:

- *Minimal Regurgitation:* This means that the amount of blood lost upstream as the valve closes is small. For example, closure regurgitation through the mitral valve would result in some blood loss from the left ventricle to the left atrium as the mitral valve closes. Some degree of valvular regurgitation is inevitable and natural (Fixme: Give indicative value). However, several heart valve pathologies (e.g. rheumatic endocarditis) may lead to clinically significant valvular regurgitation. A desirable characteristic of heart valve prostheses is that regurgitation is minimal over the full range of physiological heart function (i.e. complete functional envelope of cardiac output vs. heart rate).

- *Minimal transvalvular pressure gradient*: Whenever a fluid flows through a restriction, such as a valve, a pressure gradient arises over the restriction. This pressure gradient is a result of the increased resistance to flow through the restriction. Natural heart valves have a low transvalvular pressure gradient as they present little obstruction to the flow through themselves, normally less than 16 mmHg. A desirable characteristic of heart valve prostheses is that their transvalvular pressure gradient is as small as possible.
- *Non-thrombogenic*: As natural heart valves are lined with an endothelium continuous with the endothelium lining the heart chambers they are not normally thrombogenic. This is important as should thrombi form on the heart valve leaflets and become seeded with bacteria, so called "bacterial vegetations" will form. Such vegetations are difficult for the body to deal with as the normal physiological defence mechanisms are not present within the valve leaflets because they are avascular and largely composed of connective tissue (Fixme: Create article discussing the pathgonesis of leaflet bacterial vegetations.). Should bacterial vegetations form on the valve leafets they may continually seed bacteria into the arterial tree which may lead to bacteremia or septicaemia. Portions of the vegetation may also break off forming septic emboli. Septic emboli can lodge anywhere in the arterial tree (e.g. brain, bowel, lungs) causing local infectious foci. Even dislodged fragments from uninfected thrombi can be hazardous as they can lodge in, and block, downstream arteries (e.g. coronary arteries leading to myocardial infarction, cerebral arteries leading to stroke). A desirable characteristic of heart valve prostheses is that they are non or minimally thrombogenic.
- *Self-repairing*: Although of limited extent compared to well vascularised tissue (e.g. muscle), the valve leaflets do retain some capacity for repair due to the presence of regenerative

cells (e.g. fibroblasts) in the connective tissue from which the leaflets are composed. As the human heart beats approximately 3.4x10 times during a typical human lifespan this limited but nevertheless present repair capacity is critically important. No heart valve prostheses can currently self-repair but replacement tissues grown using stem cell technology may eventually offer such capabilities. (State that they wear).

- Rapid dynamic response - STD

Design Challenges of Heart Valve Prostheses

A replaceable model of Cardiac Biological Valve Prosthesis.

- Thrombogenesis / haemocompatibility
  - o Mechanisms:
    - — Forward and backward flow shear
    - — Static leakage shear
    - — Presence of foreign material (i.e. intrinsic coagulation cascade)
    - — Cellular maceration
- Valve-tissue interaction
- Wear
- Blockage
- Getting stuck
- Dynamic responsiveness
- Failure safety
- Valve orifice to anatomical orifice ratio
- Trans-valvular pressure gradient
- Minimal leakages
- Replaceable Models of Biological Valves.

Pulmonary Valve

The pulmonary valve (sometimes referred to as the pulmonic valve) is the semilunar valve of the heart that lies between the right ventricle and the pulmonary artery and has three cusps. Similar to the aortic valve, the pulmonary valve opens in ventricular systole, when the pressure in the right ventricle rises above the pressure in the pulmonary artery. At the end of ventricular systole, when the pressure in the right ventricle falls rapidly, the pressure in the pulmonary artery will close the pulmonary valve.

The closure of the pulmonary valve contributes the P2 component of the second heart sound (S2). The right heart is a low-pressure system, so the P2 component of the second heart sound is usually softer than the A2 component of the second heart sound. However, it is physiologically normal in some young people to hear both components separated during inhalation.

Larner-Johnson Valve

Larner-Johnson Valve is a mechanism used in dams and water pumping to control the flow of water through large water pipes. It was manufactured in the early 20th century by the Larner-Johnson Company in the USA. The valves are still manufactured in the United Kingdom by Blackhall Engineering Ltd.

In 2009, Blackhall Engineering supplied four 60" bore Larner-Johnson valves to New York City Department of Environmental Protection for the Ashokan Reservoir in upstate New York. Each valve is capable of passing a flow rate of 19 cubic metres/second, which is the equivalent of 19 tons of water per second. The valves control the flow of water out of the Ashokan Reservoir into the Catskill Aqueduct down to New York City.

Leaf Valve

A leaf valve is a one-way valve (check valve) consisting of a diagonal obstruction with an opening covered by a hinged flap.

Patents exist in the United states which specify the mechanical use of leaf valves to control various types of fluids. United States Patent 4930535 is a folding leaf valve which is described as a valve that, "folds upon itself to provide positive sealing of a pressurized chamber under a variety of pressure conditions. In particular, the valve stem folds to effectively close the lumen and provide a positive seal even at low differential pressure."

United States Patent 4795340 is a single leaf valve for a gas-fired boiler. It has a seat for a valve leaf, and a single valve leaf for the two constituents, this single valve leaf being flexible and deformable.... Application, in particular, to gas-fired boilers of the pulsed-combustion type." United States Patent 5441133 This hydraulic damper has a leaf valve stopper, the patented invention relates to a hydraulic damper. A leaf valve and a main valve are provided at an inner seat and an outer seat having different heights to each other arranged at outlet ports of a piston in such a way as they may be opened or closed... an attenuation force can be stabilised.. and the leaf valve is supported by the tapered or curved surface of the stopper when flexed, resulting in that no cracks are generated.

Pilot Valve

A pilot valve is a small valve that controls a limited-flow control feed to a separate piloted valve. Typically, this valve controls a high pressure or high flow feed. Pilot valves are useful because they allow a small and easily operated feed to control a much higher pressure or higher flow feed, which would otherwise require a much larger force to operate; indeed, this is even useful when a solenoid is used to operate the valve.

Pilot valves are often used in critical applications (i.e., emergency and safety controls) and are human-operated. They can be set up as a push-to-activate or dead man's switch.

Sleeve Valve

The sleeve valve is a type of valve mechanism for piston engines, distinct from the usual poppet valve. Sleeve-valve engines saw use in a number of pre-World War II luxury cars and in USA in the Willys-Knight car and light truck. They subsequently fell from use due to advances in poppet-valve technology, including sodium cooling, and to their tendency to burn a lot of lubricating oil or to seize due to lack of it. The Scottish Argyll company used its own, much simpler and efficient, single-sleeve system in its cars, a system which, after extensive development, saw substantial use in aircraft engines of the 1940s, such as the Napier Sabre and Bristol Hercules and Centaurus, only to be supplanted by the jet engine.

Description

A sleeve valve takes the form of one or more machined sleeves. It fits between the piston and the cylinder wall in the cylinder of an internal combustion engine where it rotates and/or slides, ports (holes) in the side of the sleeve(s) aligning with the cylinder's inlet and exhaust ports at the appropriate stages in the engine's cycle.

Types of Sleeve Valve

The first successful sleeve valve was patented by Charles Yale Knight, and used twin alternating sliding sleeves. It was used in some luxury automobiles, notably Daimler, Mercedes-Benz, Minerva and Avions Voisin. The higher oil consumption was heavily outweighed by the quietness of running and the very high mileages without servicing. Early poppet-valve systems required decarbonization at very low mileages. The Burt-McCollum sleeve valve, as used by the Scottish company Argyll for its cars, and later adopted by Bristol for its radial aircraft engines, used a single sleeve which rotated around a timing axle set at 90 degrees to the cylinder axle. Mechanically simpler and more rugged, the Burt-McCollum valve had the additional advantage of reducing oil consumption (compared to other sleeve valve designs), while retaining the rational combustion chambers and big, uncluttered, porting area possible in the Knight system.

A small number of designs used a "cuff" sleeve in the cylinder head instead of the cylinder proper, providing a more "classic" layout compared to traditional poppet valve engines. This design also had the advantage of not having the piston within the sleeve, although in practice this appears to have had little practical value. On the downside, this arrangement limited the size of the ports to that of the cylinder head, whereas in-cylinder sleeves could have much larger ports.

Advantages/Disadvantages

Advantages: The main advantages of the sleeve-valve engine are:

- Increased volumetric efficiency due to very large port openings. Sir Harry Ricardo also demonstrated better mechanical efficiency. An additional advantage of the system is that the size of the ports can be readily controlled. This is important when an engine operates over a wide RPM range, since the speed at which air can enter and exit the cylinder is defined by the size of the duct leading to the cylinder, and varies according to the cube of the RPM. In other words, at higher RPM the engine typically requires larger ports that remain open for a greater proportion of the cycle, which is fairly easy to achieve with sleeve valves, but difficult in a poppet valve system.
- Good exhaust scavenging and controllable swirl of the inlet air/fuel mixture in single-sleeve designs. When the intake ports open, the fuel air mixture can be made to enter tangentially to the cylinder. This helps scavenging when exhaust/inlet timing overlap is used and a wide speed range required, whereas poor poppet valve exhaust scavenging can dilute the fresh air/fuel mixture intake to a greater degree, being more speed dependent (relying principally on exhaust/inlet system resonant tuning to separate the two streams). Greater freedom of combustion

chamber design (few constraints other than the spark plug positioning) means that fuel/air mixture swirl at TDC can also be more controlled allowing improved ignition and flame travel which as demonstrated by Ricardo, at least one extra unit of compression ratio before detonation c.f. the poppet valve engine.

- The combustion chamber formed with the sleeve at the top of its stroke is ideal for complete, detonation-free combustion of the charge, as it does not have to contend with compromised chamber shape and hot exhaust (poppet) valve(s).
- No springs are involved in the sleeve valve system, therefore the power needed to operate the valve remains largely constant with the engine's RPM, meaning that the system can be used at very high speeds with no penalty for doing so. A problem with high-speed engines which use poppet valves is that as engine speed increases, the speed at which the valve moves also has to increase. This in turn increases the loads involved due to the inertia of the valve, which has to be opened quickly, brought to a stop, then reversed in direction and closed and brought to a stop again. Large valves that allow good air-flow have considerable mass and require a strong spring to overcome the opening inertia. At some point, the valve spring reaches its resonance frequency, causing a compression wave to oscillate within the spring, which in turn causes it to become effectively weaker and unable to properly close the valve. This valve float can result in the valve not closing quickly, and it may strike the top of the rising piston. In addition, camshaft, pushrods, and valve rockers can be eliminated in a sleeve valve design, as the sleeve valves are generally driven by a single gear powered from the crankshaft. In an aircraft engine this provided reductions in weight and complexity.
- Longevity, as demonstrated in early automotive applications of the Knight engine. Prior to the advent of leaded gasolines, poppet-valve engines typically required grinding of the valves and valve seats after 20,000 to 30,000 miles (32,000 to 48,000 km) of service. Sleeve valves did not suffer from the wear and recession caused by the repetitive impact of the poppet valve against its seat. Sleeve valves were also subjected to less intense heat buildup than poppet valves, owing to their greater area of contact with other metal surfaces. In the Knight engine, carbon build-up actually helped to improve the sealing of the sleeves, the engines being said to "improve with use",

in contrast to poppet valve engines, which lose compression and power as valves and valve stems/guides wear. Due to the continued motion of the sleeve (Burt-McCollum type), the high wear points linked to poor lubrication in the TDC/BDC of piston travel within the cylinder are suppressed, therefore rings and cylinders lasted much longer.

- Cylinder head is not required to house valves, allowing the spark plug to be placed in the best possible location for efficient ignition of the combustion mixture. For very big engines, where flame propagation speed limits both size and speed, the swirl induced by ports as described by Ricardo can be an additional advantage.
- Continental in the USA conducted extensive research in Single-Sleeve-Valve engines, pointing that they were eventually of lower production cost, and easier to produce, however, their aircraft engines soon equalled the SSV engines performance by introducing improvements such as Sodium cooled poppet valves, and it seems also that the costs of this research, along with the October 1929 crisis, lead to the Continental SSV engines not entering mass production. (Continental ! Its Motors and its People. William Wagner, 1983, p. 33). The same book reports that General Motors had made tests with SSV engines, rejecting this kind of arrangement.
- If left on a horizontal surface, sleeves tend to ovalize, producing several types of mechanical problems. In order of avoiding this, special cabinets were built, where sleeves were kept in a vertical position, inserted in a cylinder of appropriate bore and height.

Most of these advantages were evaluated and established during the 1920s by Sir Harry Ricardo, possibly the sleeve-valve engine's greatest advocate. He conceded that some of these advantages were significantly eroded as fuels improved up to and during World War II and as sodium-cooled exhaust valves were introduced in high output aircraft engines.

Disadvantages

The sleeve valve's one major disadvantage is that perfect sealing is difficult to achieve. In a poppet valve engine, the piston possesses piston rings (often at least three and sometimes as many as eight) which form a seal with the cylinder bore. During the "breaking in" period (known as "running-in" in the UK) any imperfections in one are scraped into the other, resulting in a good fit. This type of "breaking in" is not possible on a sleeve-valve engine, however, because the piston and sleeve move in different directions and in some systems even rotate in relation to one another. Unlike a traditional design, the imperfections in the piston do not always line up with the same point on the sleeve. In the 1940s this was not a major concern because the poppet valves of the time typically leaked appreciably more than they do today, so that oil consumption was significant in either case. One of the 1911–1928 Argyll Single Sleeve Valve engines, the 15/30, a four-cylinder 2'610 cc unit, was attributed an oil consumption of one gallon for 1'000 (one thousand) miles. Mike Hewland in 1974 claimed that the progresses in lubricating oils, materials, and machining had solved the oil thirst problem, you can see some modern homemade SSV engines being a little bit smoky sometimes during its operation.

The high oil consumption problem associated with the Knight double sleeve valve was fixed with the Burt-McCollum single sleeve valve, as perfected by Bristol. At top dead centre (TDC), the single sleeve valve rotates in relation to the piston. This prevents boundary lubrication problems, as piston ring ridge wear at TDC and bottom dead centre (BDC) does not occur. The Bristol Hercules overhaul life was rated at 3,000 hr. An inherent disadvantage may be that the piston in its course partially obscures the ports, thus making it difficult for gases to flow during the crucial overlap between the intake and exhaust valve timing usual in modern engines. The German engineer Max Bentele, after studying a British sleeve valve aero engine (probably a Hercules), complained that the arrangement required more than 100 gearwheels for the engine, too many for his taste.

History

Charles Yale Knight: In 1901 Knight bought an air-cooled, single-cylinder three-wheeler whose noisy valves annoyed him. He believed that he could design a better engine and did so, inventing his double sleeve principle in 1904. Backed by Chicago entrepreneur L.B. Kilbourne, a number of engines were constructed followed by the "Silent Knight" touring car which was shown at the 1906 Chicago Auto Show.

Knight's design had two cast-iron sleeves per cylinder, one sliding inside the other with the piston inside the inner sleeve. The sleeves were operated by small connected rods actuated by an eccentric shaft. They had ports cut out at their upper ends. The design was remarkably quiet, and the sleeve valves needed little attention. It was, however, more expensive to manufacture due to the precision grinding required on the sleeves' surfaces. It also used more oil at high speeds and was harder to start in cold weather.

Although he was initially unable to sell his Knight Engine in the United States, a long sojourn in England, involving extensive further development and refinement by Daimler supervised by their consultant Dr Frederick Lanchester eventually secured Daimler and several luxury car firms as customers willing to pay his expensive premiums. He first patented the design in England in 1908. As part of the licensing agreement, "Knight" was to be included in the car's name.

Among the companies using Knight's technology were Gabriel Voisin (in his Avions Voisin cars), Daimler (even in their V-12 "Double Six", from 1909–1930s), Panhard (1911–39), Mercedes (1909–24), Willys (as the Willys-Knight, plus the associated Falcon-Knight), Stearns, Mors, Peugeot, and Belgium's Minerva company, some thirty companies in all. Itala also experimented with sleeve valves.

Upon Knight's return to America he was able to get some firms to use his design; here his brand name was "Silent Knight" (1905–1907) — the selling point was that his engines were quieter than those with standard poppet valves. The best known of these were the F.B. Stearns Company of Cleveland, which sold a car named the Stearns-Knight, and the Willys firm which offered a car called the Willys-Knight, which was produced in far greater numbers than any other sleeve-valve car.

Burt-McCollum

The Burt-McCollum sleeve valve consisted of a single sleeve, which was given a combination of up-and-down and partial rotary motion. It was developed in about 1909 and was first used in the 1911 Argyll car. Argyll went out of business after high expenses of a litigation with the Knight patent holders. Its greatest success was in Bristol's large aircraft engines, and was also used in the Napier Sabre and Rolls-Royce Eagle aircraft engines. The single sleeve system also cured the high oil consumption associated with the Knight double sleeve valve.

A number of sleeve valve aircraft engines were developed following a seminal 1927 research paper from the RAE by Harry Ricardo. This paper outlined the advantages of the sleeve valve, and suggested that poppet valve engines would not be able to offer power outputs much beyond 1500 hp (1,100 kW). Napier and Bristol began the development of sleeve-valve engines that would eventually result in two of the most powerful piston engines in the world: the Napier Sabre and Bristol Centaurus.

Potentially the most powerful of all sleeve-valve engines (though it never reached production) was the Rolls-Royce Crecy V-12 (oddly, using a 90-degree V-angle), two-stroke, direct-injected, force-scavenged (turbocharged) aero-engine of 26.1 litres capacity. It achieved a very high specific output, and surprisingly good specific fuel consumption (SFC). In 1945 the single-cylinder test-engine (E65) produced the equivalent of 5,000 HP (192 BHP/Litre) when water injected, although the full V12 would probably have been initially type rated at circa 2,500 hp (1,900 kW). Sir Harry Ricardo, who specified the layout and design goals, felt that a reliable 4,000 HP military rating would be possible. Ricardo was constantly frustrated during the war with Rolls-Royce's (RR) efforts. Hives & RR were very much focused on their Merlin, Griffon, then Eagle and finally Whittle's jets, which had a clearly defined production purpose. Ricardo and Tizard eventually realised that the Crecy would never get the development attention it deserved unless it was specified for installation in a particular aircraft, but by 1945, their "Spitfire on steroids" concept of a rapidly climbing interceptor powered by the lightweight Crecy engine had become an aircraft without a purpose.

Following World War II the sleeve valve disappeared from use, as the previous problems with sealing and wear on poppet valves had been remedied by the use of better materials, and the inertia problems with the use of large valves were reduced by using several smaller valves instead, giving increased flow area and reduced mass. Up to that point, the single sleeve valve had won every contest against the poppet valve hands down in comparison of power to displacement. The difficulty of nitride hardening, then finish-grinding the sleeve valve for truing the circularity, may have been a factor in its lack of commercial application.

Modern Usage

The sleeve valve has begun to make something of a comeback, due to modern materials, dramatically better engineering tolerances and modern construction techniques, which produce a sleeve valve that leaks very little oil. However, most advanced engine research is concentrated on improving other internal combustion engine designs, such as the Wankel.

Mike Hewland with his assistant John Logan, and Keith Duckworth experimented with a single-cylinder sleeve-valve test engine when looking at Cosworth DFV replacements. Hewland claimed to have obtained 72 hp (54 kW) from a 500 cc single-cylinder engine, with a specific fuel consumption of 170 gr/HP/hr -.45 to .39 lb/hp/hr-, the engine being able to work on creosote, with no specific lubrication supply for the sleeve, just the sleeve driving mechanism requires an specially devoted lubrication. Hewland reported also that the highest temperature measured in the cylinder head didn't exceed 150 °C, sleeve temperatures were around 140 °C, T was 270 °C in the centre of cylinder and 240 °C in the edge.

A recent SAE paper deals with a high-speed, small-displacement sleeve-valve engine, calculated, but not experimentally shown, to have a higher SFC than the poppet-valve alternative, a non-surprising result, considering the difficulty in obtaining the high intake and exhaust overlap that very fast-running engines require, additional work compares two different side-opening intake strategies for sleeve-valve engines. An unusual form of four-stroke model engine that uses what is essentially a sleeve-valve format, is the British RCV series of "SP" model engines, which use a rotating cylinder liner driven through a bevel gear at the cylinder liner's "bottom", and even more unusually have the propeller shaft - as an integrally machined part of the rotating cylinder liner - emerging from what would normally be the cylinder's "top", at the extreme front of the engine, achieving a 2:1 gear reduction ratio compared to the vertically oriented crankshaft's rotational speed. The same firm's "CD" series of model engines use a conventional upright single cylinder instead, with the crankshaft used to directly spin the propeller, and also use the rotating cylinder valve. As a parallel with the earlier Charles Knight-designed sleeve-valved automotive powerplants, any RCV sleeve-valved model engine that is run on model glow engine fuel using castor oil as a small percentage (about 2% to 4% content) of the lubricant in the fuel allows the "varnish" created through engine operation to provide a better pneumatic seal between the rotating cylinder valve and the unitized engine cylinder/head castings, initially formed while the engine is being broken in.

Steam Engine

Sleeve valves have occasionally been used on steam engines, for example the SR Leader class.

Schrader Valve

The Schrader valve (also called American valve) is a brand of pneumatic tire valve used on virtually every motor vehicle in the world today. The Schrader company, for which it was named, was founded in 1844 by August Schrader. The original Schrader valve design was patented in the United States in 1893.

The Schrader valve consists of a valve stem into which a valve core is threaded, and is used on virtually all automobile tires and most wider rimmed bicycle tires. The valve core is a poppet valve assisted by a spring.

Uses: In addition to tube and tubeless tires, Schrader valves of varying diametres are used on many refrigeration and air conditioning systems to allow servicing, including recharging with refrigerant; by plumbers conducting leak-down pressure tests on pipe installations; as a bleeding and test port on the fuel rail of some fuel injected engines; and in the buoyancy compensator (BC) inflators of SCUBA systems (the inflation button pushes down on the valve, allowing pressurized gas to flow into the BC). Schrader valves are also widely used in high-pressure hydraulic systems on aircraft. Many domestic fire extinguishers use an internal valve identical to a Schrader valve, but with a lever on top to enable quick release of the pressurised content.

Valve

A Schrader valve consists of an externally threaded hollow cylindrical metal tube, typically of brass. In the centre of the exterior end is a metal pin pointing along the axis of the tube; the pin's end is approximately flush with the end of the valve body.

Generally, all Schrader valves used on tires have threads and bodies of a single standard size at the exterior end, so caps and tools generally are universal for the valves on all common applications. The core of the valve can be removed or tightened with a tool.

A new development is Schrader valve stems with integrated transmitters for tire-pressure monitoring systems (TPMS).

Cap

A valve cap is important on a Schrader valve because if one is not fitted, dirt and water can enter the outside of the valve, potentially jamming it or contaminating the sealing surfaces and causing a leak. Rock salt and other chemical de-icers used in the winter are especially damaging for the brass components in the Schrader valve.

Some valve caps have a rubber-washer seal (or a synthetic-rubber-washer seal) inside to help make a hermetic seal (airtight seal). A such rubber-washer-sealing-type cap also helps prevent air from escaping from a slightly leaking valve. In addition, the rubber washer prevents the cap from loosening and falling off due to vibrations and hence being lost, by acting as a cushion between the cap and the valve stem and in this way damping the vibrations. The caps without these rubber-washer seals inside them effectively serve mostly to keep contaminants out of the valve stem, without providing a good hermetic seal. The caps without rubber washers inside them can also loosen and fall off due to vibrations when undertightened, especially if they are made of hard plastic or hard-plastic-lined metal. Valve caps are mostly made of black plastic and also of metal or plastic-lined metal, with rubber washers found inside some caps made of plastic and most metal caps. However, to prevent corrosion due to electrolysis and malfunction of the electronics in a direct TPMS (tire-pressure-monitoring system) and avoid costly repairs, metal valve caps without insulating plastic linings should never be used on direct-TMPS-equipped tire valve stems. Therefore, only plastic caps or metal caps with plastic linings should be used in direct-TMPS-equipped tire valve stems. Moreover, metal valve caps without plastic linings can also cause corrosion and may become stuck to the valve and require breaking of the valve in order to remove it when the cap metal (such as brass) is different than the valve metal (such as aluminium). When the valve caps are installed, care must be made not to leave them loose, as they may not seal properly and may even come loose and fall off. Overtightening the valve caps might damage the rubber washer inside the cap, damage or break the cap itself, or damage the valve. Also, installing the cap at a slight angle can cause stripping of the threads and, therefore, it should be gently finger-tightened while ensuring that it is straight and the threads aren't stripped.

There are also special pressure monitoring valve caps available that use a spring-loaded piston to raise a green flag when the pressure is at or above the correct setting. Upon losing pressure the green flag is retracted to reveal a red pin, hopefully catching the attention of the owner before fuel is wasted by running the tire under-inflated.

Recently, coloured plastic valve caps have appeared. Certain automobile tire dealerships are promoting the use of dry nitrogen to inflate tires. Dealers claim that eliminating oxygen and water will prolong the life of both tires and wheels. These dealers install green caps to signify that the tires are filled with nearly pure (typically about 95%) nitrogen. In contrast, regular air is approximately 78% nitrogen. The moisture in the air-filled tires condenses in below-freezing temperatures and causes a sudden drop in tire pressure. Dry-nitrogen filling helps prevent this problem. As an alternative to green nitrogen-ID (nitrogen-identification) caps, green nitrogen-ID O-rings to be installed on valve stems under the caps, while still using black- or metallic-coloured caps, are also available.

Other vendors are selling caps in a variety of other colours for purely decorative purposes. The decorative category even includes caps that light up when the wheels move. When used on a bicycle these can also enhance safety by improving the visibility of the rider.

Presta Valves vs. Schrader Valves

Whereas Schrader valves are almost universal on car tires, bicycle tubes have Dunlop, Schrader or Presta valves. Both types are good at sealing high pressures. Their chief differences are that Schrader valves are larger and have springs that close the valve except when the pin is depressed. Schrader valves are used in a wide variety of compressed gas and compressed liquid applications.

Compared to Presta valves, Schrader valves used for bicycle tires have a greater diametre, requiring a larger diametre hole in a bicycle rim. While not a concern on wider bike rims, it will weaken a narrow wheel rim, precluding their use on (road) racing bicycles. Another disadvantage of the Schrader is that the air chuck must depress the spring-loaded pin before air can flow during inflation, whereas the Presta valve relies solely on internal air pressure to keep it shut. This means that some air is lost while attaching and detaching pumps to Schrader valves, although this is usually nominal (most of the air which makes the "whooshing" noise when the pump is removed comes from the compressed air in the tube between the pump body and the attachment end, not the tire via the momentarily-open Schrader valve). To fill up a bicycle tire with a Presta-valved tube at a gas station requires an adaptor, while a Schrader-valved tube does not.

Dimensions

Schrader valves are classified by their material, diametre of intended rim hole, length, and shape.

- TR-4 - straight metal stem (8 mm dia.)
- TR-6 - straight metal stem (8 mm dia.)
- TR-13 - straight rubber stem (11.5 mm dia.)
- TR-15 - straight rubber stem (16 mm dia.)
- TR-87 - short 90° metal stem (10 mm dia.)
- TR-87C - tall 90° metal stem (10 mm dia.)

And also :

- TR 413- 0.453" rim hole diam.(11,50mm) 1,25" long (=31,8mm)
- TR 415- 0.625" rim hole diam.(15,87mm) 1,25" long (=31,8mm)
- TR 418- 0.453" rim hole diam.(11,50mm) 2" long (=50,8mm)
- TR 425- 0.625" rim hole diam.(15,87mm) 2" long (=50,8mm)

The standard Schrader valve has the following threads:

External Thread :

- *Metric*: 7.7 mm OD, thread root diametre is 6.9 mm × 0.794 mm pitch.
- *Imperial*: 0.305 in OD, thread root diametre 0.271 in × 32 tpi (threads per inch)

Internal thread (to accept the threaded valve core)

- *Metric*: 5.30 mm OD × 0.706 mm pitch
- *Imperial*: 0.209 in OD × 36 tpi.

For refrigeration, a 1/4" male flare fitting is used, with the same internal thread as above.

Presta Valve

The Presta valve (also called Sclaverand valve or French valve) is a valve commonly found in high pressure road style and many mountain bicycle inner tubes. It comprises an outer valve stem and an inner valve body. A lock nut to secure the stem at the wheel rim and a valve cap may also be present.

The outer valve stem is manufactured in various lengths for different applications, and has a narrower diametre (6 mm) than the Schrader valve (8 mm), which helps strengthen narrow rims because the weakest point of a bicycle rim is usually the valve hole.

The air pressure in an inflated tire holds the inner valve body shut. A small screw and captive nut on the top of the valve body permits the valve to be screwed shut and ensure that it remains tightly closed. The nut must be unscrewed to permit airflow in either direction (this must be done before attaching a pump). The screw remains captive on the valve body even when unscrewed fully; it is tightened again after the tire is inflated and the pump removed.

The valve cap protects the valve body, and also prevents the valve from damaging the tube when it is rolled for storage, but is not necessary to prevent pressure loss. Because the rims of bicycles drilled for Presta valves cannot accommodate the wider Schrader valves, it is often the case that rims need to be drilled for such replacements. Conversely, when a Presta valve is fitted into the larger Schrader rim hole, grommets or reducers are sometimes used to take up the extra space. The standard Presta valve has an external thread. An adaptor can be fitted onto this external thread to permit the Presta valve to be connected to a pump with a Schrader chuck. The same adaptor, because of a coincidence of thread sizes, can convert a direct-fitting Schrader pump into one that can connect to flexible adaptors of either kind.

Tubeless Systems

Unlike Schrader cores, the inner parts of the valve, the valve core, of standard Presta valves often cannot be removed from the stem. Since the core usually has to be removed to use a tubeless or Universal System Tubeless (UST) setup so that tire sealant may be added, removable core Presta valves have become more common.

Valve Stem

A valve stem is a self-contained valve which opens to admit gas to a chamber (such as air to inflate a tire), and is then automatically closed and kept sealed by the pressure in the chamber, or a spring, or both, to prevent the gas from escaping. They are most commonly used on automobile and bicycle wheels, but also for many other applications.

Schrader: Schrader valves (also called American valves) consist of a valve stem into which a poppet valve is threaded with a spring attached. They are used on virtually all automobile tires, and in the United States, on most wider rimmed bicycle tires. In addition to tires, Schrader valves of varying diametres are used in refrigeration and air conditioning systems, plumbing, fuel injection engines, suspension systems, and SCUBA regulators, allowing the user to remove and attach the hose while in use.

Presta

Presta valves (also called Sclaverand valves or French valves) are normally only used on bicycles. The stem has a narrower diametre (nominally 6 mm) than the thinnest (nominally 8 mm) Schrader type, and so the bore of the hole in the rim through which the stem passes can be smaller.

Dunlop (or Woods)

Dunlop valves, (also called Woods valves or English valves) were once popular in Great Britain. They are still widely used on bicycle tires in many countries (Japan, India, Russia, the Netherlands, Germany, Finland, Sweden, and developing countries), especially on low to medium priced bicycles.

Others

Many other valves exist that are used only in certain regions or for limited purposes.

For example the "Regina Valve" is a valve very much similar to the Presta and mostly being used in Italy. Invisible tire valves are different in design from traditional tire valves. An inflator stem is not used; instead, a removable cap on a valve stem embedded in the tire rim is fitted, with only the cap visible. When inflating the tire, the cap is first removed, typically with a coin, and then a "portable" dedicated inflator stem is screwed onto the valve stem. Through the inflator the tire is inflated as usual.

Dunlop Valve

The Dunlop valve, (also called a Woods valve or an English valve) is a type of pneumatic valve stem in use in some countries, such as Japan, Czech Republic, Poland, Russia, the Netherlands, Germany, Britain, Finland, Sweden, and developing countries. It has a wider base than a Presta valve, similar in size to a Schrader valve, but it can be inflated with Presta valve adapter. The inner mechanism of the valve can be replaced easily, without the need for special tools.

Rupture Disc

A rupture disc, also known as a bursting disc or burst diaphragm, is a non-reclosing pressure relief device that, in most uses, protects a pressure vessel, equipment or system from overpressurization or potentially damaging vacuum conditions. A rupture disc is a type of sacrificial part because it has a one-time-use membrane that fails at a predetermined differential pressure, either positive or vacuum. The membrane is usually made out of metal, but nearly any material (or different materials in layers) can be used to suit a particular application.

Rupture discs provide instant response (within milliseconds) to an increase or decrease in system pressure, but once the disc has ruptured it will not reseal. Major advantages of the application of rupture discs compared to using pressure relief valves include leak-tightness and cost. Rupture discs are commonly used in petrochemical, aerospace, aviation, defence, medical, railroad, nuclear, chemical, pharmaceutical, food processing and oilfield applications. They can be used as single protection devices or as a backup device for a conventional safety valve; if the pressure increases and the safety valve fails to operate (or can't relieve enough pressure fast enough), the rupture disc will burst. Rupture discs are very often used in combination with safety relief valves, isolating the valves from the process, thereby saving on valve maintenance and creating a leak-tight pressure relief solution.

Some models of gene gun also use a rupture disk, but not as a safety device. Instead, their function is part of the normal operation of the device, allowing for precise pressure-based control of particle application to a sample. In these devices, the rupture disk is designed to fail within a optimal range of gas pressure that has been empirically associated with successful particle integration into tissue or cell culture. Different disk strengths can be available for some gene gun models.

Although commonly manufactured in disc form, the devices also are manufactured as rectangular panels (rupture panels or vent panels). Device sizes range from under 1/4" to 3' or more, depending upon the industry application. Rupture discs and vent panels are constructed from carbon steel, stainless steel, hastelloy, graphite, and other materials, as required by the specific use environment.

Rupture discs are widely accepted throughout industry and specified in most global pressure equipment design codes (ASME, PED, etc.). Rupture discs can be used to specifically protect installations against unacceptably high pressures or can be designed to act as one-time valves or triggering devices to initiate with high reliability and speed a sequence of actions required.

Blowout Panel

Blowout panels, also called blow-off panels, areas with intentionally weakened structure, are used in enclosures, buildings or vehicles where a sudden overpressure may occur. By failing in a predictable manner, they channel the overpressure or pressure wave in a direction where it causes controlled, directed minimal harm, instead of causing a catastrophic failure of the structure. Blow-off panels are used in ammunition compartments of some tanks to protect the crew in case of ammunition explosion, turning a catastrophic kill into mere firepower kill. An alternate example is a deliberately weakened wall in a room used to store compressed gas cylinders; in the event of a fire or other accident, the tremendous energy stored in the (possibly flammable) compressed gas is directed into a "safe" direction, rather than potentially collapsing the structure in a similar manner to a thermobaric weapon.

Blowout panels are installed in several modern tanks, including the M1 Abrams and T-80, and have in the past been considered as a possible solution to magazine explosions on battleships.

Saddle Valve

A saddle valve is a valve used to supply liquid where a low volume, low pressure stream is required. The name is derived from the fact that it is mounted in such a way that it "saddles" the line, mounting it from both sides.

They are typically used for supplying cold water, via a 3/8 inch tube, to:

- Humidifiers and
- Icemakers in freezers.

A saddle valve is mounted directly on to a pipe which is usually 1/2 inch copper tubing. Saddle valves are self-tapping devices. Once mounted on a line, with the included rubber seal in place, all that is required is to turn the valve clockwise until it pierces the water line. When the valve handle can no longer be turned, it is done tapping the pipe. Turning the handle counter-clockwise opens the valve. These valves can be installed without shutting off the main water supply.

Plumbing codes, local or international, often do not allow use of a saddle valve; instead a tee and a conventional globe, gate, or ball valve is to be installed. However, where allowed, saddle valves are low cost and easy to install.

Solenoid Valve

A solenoid valve is an electromechanical valve for use with liquid or gas. The valve is controlled by an electric current through a solenoid: in the case of a two-port valve the flow is switched on or off; in the case of a three-port valve, the outflow is switched between the two outlet ports. Multiple solenoid valves can be placed together on a manifold. Solenoid valves are the most frequently used control elements in fluidics. Their tasks are to shut off, release, dose, distribute or mix fluids. They are found in many application areas. Solenoids offer fast and safe switching, high reliability, long service life, good medium compatibility of the materials used, low control power and compact design. Besides the plunger-type actuator which is used most frequently, pivoted-armature actuators and rocker actuators are also used. A solenoid valve has two main parts: the solenoid and the valve. The solenoid converts electrical energy into mechanical energy which, in turn, opens or closes the valve mechanically. A *direct acting* valve has only a small flow circuit, shown within section E of this diagram (this section is mentioned below as a pilot valve).

This *diaphragm piloted valve* multiplies this small flow by using it to control the flow through a much larger orifice. Solenoid valves may use metal seals or rubber seals, and may also have electrical interfaces to allow for easy control. A spring may be used to hold the valve opened or closed while the valve is not activated.

The diagram to the right shows the design of a basic valve. At the top figure is the valve in its closed state. The water under pressure enters at A. B is an elastic diaphragm and above it is a weak spring pushing it down. The function of this spring is irrelevant for now as the valve would stay closed even without it. The diaphragm has a pinhole through its centre which allows a very small amount of water to flow through it. This water fills the cavity C on the other side of the diaphragm so that pressure is equal on both sides of the diaphragm, however the compressed spring supplies a net downward force. The spring is weak and is only able to close the inlet because water pressure is equalised on both sides of the diaphram.

In the previous configuration the small conduit D was blocked by a pin which is the armature of the solenoid E and which is pushed down by a spring. If the solenoid is activated by drawing the pin upwards via magnetic force from the solenoid current, the water in chamber C will flow through this conduit D to the output side of the valve. The pressure in chamber C will drop and the incoming pressure will lift the diaphragm thus opening the main valve. Water now flows directly from A to F.

From this explanation it can be seen that this type of valve relies on a differential of pressure between input and output as the pressure at the input must always be greater than the pressure at the output for it to work. Should the pressure at the output, for any reason, rise above that of the input then the valve would open regardless of the state of the solenoid and pilot valve.

In some solenoid valves the solenoid acts directly on the main valve. Others use a small, complete solenoid valve, known as a pilot, to actuate a larger valve. While the second type is actually a solenoid valve combined with a pneumatically actuated valve, they are sold and packaged as a single unit referred to as a solenoid valve. Piloted valves require much less power to control, but they are noticeably slower. Piloted solenoids usually need full power at all times to open and stay open, where a direct acting solenoid may only need full power for a short period of time to open it, and only low power to hold it.

Types: Many variations are possible on the basic, one way, one solenoid valve described above:

- one or two solenoid valves;
- direct current or alternating current powered;
- different number of ways and positions;

Common Uses

Solenoid valves are used in fluid power pneumatic and hydraulic systems, to control cylinders, fluid power motors or larger industrial valves. Automatic irrigation sprinkler systems also use solenoid valves with an automatic controller. Domestic washing machines and dishwashers use solenoid valves to control water entry to the machine. In the paintball industry, solenoid valves are usually referred to simply as "solenoids." They are commonly used to control a larger valve used to control the propellant (usually compressed air or CO_2). Solenoid valves are used in dental chairs to control air flow. In the industry, "solenoid" may also refer to an electromechanical solenoid commonly used to actuate a sear.

Besides controlling the flow of air and fluids solenoids are used in pharmacology experiments, especially for patch-clamp, which can control the application of agonist or antagonist.

Chapter 3

Thermostatic Mixing Valve

A thermostatic mixing valve (TMV) is a valve that blends hot water with cold water to ensure constant, safe shower and bath outlet temperatures, preventing scalding. The storage of water at high temperature removes one possible breeding ground for Legionella; the use of a thermostat, rather than a static mixing valve, provides increased safety against scalding, and increased user comfort, because the hot-water temperature remains constant.

Many TMVs use a wax thermostat for regulation. They also shut off rapidly in the event of a hot or cold supply failure to prevent scalding or thermal shock. It is increasingly common practice around the world to regulate the storage water temperature to above 60 °C (140 °F), and to circulate or distribute water at a temperature less than 50 °C (122 °F). Water above these temperatures can cause scald injuries. Many countries, states, or municipalities now require that the temperature of all bath water in new build and extensively refurbished domestic properties be controlled to a maximum of 48 °C (118 °F). Installing thermostatic mixing valves can ensure that water is delivered at the required temperature, thereby reducing the risk of scalding accidents; it also reduces hot water consumption from a supply that is maintained at a higher temperature.

There are three main categories for water temperature controlling devices: Heat Source, Group Control, and Point-of-Use.

Heat Source: These are used with central heating systems that use water as a medium.

- Tempering valves for use on hot water heat distribution systems
- High flow rates suitable for use in under floor (radiant) heating applications
- Allows water to be stored at a higher temperature

Group Control: These provide a uniform distribution temperature for all hot water outlets in a household.

- Designed for multi-point applications
- High flow rates (from 14 gpm to 51 gpm at 45 psi)
- Temperature stability

Point-of-Use: These are single Outlet Thermostatic Mixing Valves, often called "thermostatic faucets", "thermostat taps" or "thermostat valves".

- Designed for single point applications, such as individual showering, hand wash basin mixers, bath or tub fillers
- High level protection against scalding and thermal shock

Although other temperature regulating valves exist, thermostatic mixing valves are the preferred type in health care facilities, as they limit maximum outlet temperature, regardless of pressure or flow.

Thermostatic Radiator Valve

A Thermostatic Radiator Valve (TRV) is a self-regulating valve fitted to hot water heating system radiators. The TRV controls the temperature of a room by regulating the flow of hot water to the radiator. Thermostatic radiator valves (air vent valves) also exist for steam radiators.

Short History of the TRV

Many years ago the first ideas for heating controls saw the light of day, but it was not until 1943 when Mads Clausen, founder of Danfoss, invented the first radiator thermostat that the development of TRVs became serious. However, from the first invention to mass production to the European market it took almost 15 years, and it was not until 1973 when the first oil crisis hit the world that the installation of TRVs really took speed. Today TRVs have a wide market distribution and acceptance worldwide as an energy efficient and competitive technology. In spite of this there is still a huge potential for further implementation of TRVs as replacement of old manual valves to benefit the environment and reduce the energy costs for people around the world.

Product Design and Functionality

A TRV consists of two parts: a valve that opens or closes to control the hot water flow and a sensor that controls the opening of the valve. The sensor contains an actuator with a sensing substance, which adjusts the valve opening based on the temperature in the room and via a physical connection between the actuator-spindle and the valve-spindle/cone.

The sensors are for the most part made of plastic in many different designs and shapes. The actuator usually contains a sensing substance, such as wax, liquid or gas.

The valves are mainly made by brass or bronze and have different sizes, shapes and connections to the radiator. The majority of valves are mounted on the piping connected to the radiator; however, the valves may also be mounted as an integrated part of the radiator.

Operation

TRVs are self-regulating devices, which control the temperature in the room based on an individually set temperature and heat load. TRVs can work together with outdoor temperature controls, supply flow temperature controls, pressure controls and time set-back devices.

The overall principle can be described in this way:

1. The actuating device in the sensor is a steel container containing a sensing substance, either wax, liquid or gas. All three substances expand or contract depending on the temperature in the room.
2. A spindle system in the sensor transfers the axial movement from the actuator to the valve by interaction between the sensor spindle and the valve spindle. In the valve a rubber cone fixed on the valve spindle interacts with the valve seat and thereby regulates and controls the flow of hot water through the valve to the radiator inlet.

The process is completely self-contained and without complex electronic controls. The TRV keeps the room temperature at a desired level even through fairly wide swings in indoor temperature due to external conditions such as sudden temperature drops, solar radiation, wind velocity or wind direction. Note that a TRV cannot manage room temperature without a thermostat somewhere in the system to signal the heating source to kick in and produce hot water or steam. For example, in a three bedroom apartment with a living and dining area, there has to be a thermostat in the apartment to signal the boiler to produce hot water. If the thermostat for the apartment is located in say, the living area, then any TRV's installed on radiators in the bedrooms can manage temperatures in those rooms, provided the thermostat in the living room is calling for heat. So if we wish to turn down the thermostat in the living room when we go to bed then the bedroom radiators will only get heat if the living room temperature is above the temperature in the living room.

One way to get around this problem is to use a second, or third or fourth, thermostat in other rooms, each with a separate connection to the boiler or heating source.

After installing TRVs, owners have to get used to the self-regulating principle which means that they do not have to turn the thermostat up and down. When room temperatures vary, the valve automatically opens or closes to regulate the flow of hot water until the desired temperature is achieved. Only if the desired room temperature has to be changed or if windows are opened or closed for ventilation does the sensor setting need to be adjusted.

A TRV should not be fitted on a radiator where it is in the same room as the main house/rooms thermostat, as this can lead to inaccurate overall house temperature control. Refer to BS5449 or equivalent.

Applications and CO_2 Savings

TRVs are suitable for all kind of radiator systems, radiators, convectors and towel dryers. Designs of the sensors can be tailored to suit the specific applications. TRVs are cheap and easy solutions with a significant potential for energy savings and CO_2 reductions from heating installations. In fact the replacement of a manual heating control with a TRV can save at least 280 kg CO_2 per year.

When dimensioning a heating system in a building, it is necessary to perform a precise calculation of the total system. As part of this calculation it is important to secure proper hydraulic balancing of the complete heating system. For this balancing, the presetting feature in the valve or a special hydraulic balancing valve can be used.

Trap Primer

A trap primer (or trap seal primer) is a plumbing device or valve that adds water to traps. The water seals in traps are needed to prevent sewer gases from entering buildings, but because this water is exposed to the air, it is subject to evaporation over time in infrequently used floor drains, leading to the release of sewer gas into the environment. The trap primer mitigates this problem by injecting water, either directly or indirectly, into the trap to maintain the water seal indefinitely.

Building codes may require trap primers for some traps, especially for where experience has shown that there are isolated and/or infrequently used drains such as basement floor drains (which are only used when the basement is flooded). Where explosion hazards are high, specifying trap primers, whether required or not by code, is wise. Trap primers do require maintenance and replacement, so they should be installed in accessible locations.

Types: There are many types of trap primers. The simplest, typically for a floor drain's trap, is simply a connection from a nearby sink's drain so that when the sink is used, some of the water flow is diverted into one or more traps. More common is a primer that is connected to the potable water supply and activates when pressure fluctuations are sensed, such as the flushing of a nearby water closet. Others depend on occupancy sensors or timers. Several manufacturers (e.g., Moen) produce laundry faucets with a built-in trap seal primer outlet.

Other Solutions

In infrequently used floor drains where such a primer does not exist, evaporation can be mitigated a few different ways:

- Routinely adding water to the floor drain
- Pouring a small amount of vegetable or mineral oil into the drain to just cover the trap's water surface. The oil floats on the surface of the water and creates a liquid seal preventing further evaporation. Mineral oil tends to be a better option, as it does not go rancid, where vegetable oil does. Any environmental impact from its discharge is negligible compared to the contents that can be flushed out of a flooded basement.
- Where acceptable by local code, inline trap sealer (i.e. Canplas SureSeal) can be installed.

Poor venting or windy conditions can cause water seals to be siphoned or blown out; for these cases, the drain must be refilled.

Heart Valve Dynamics: In general, motion of the heart valves is determined using the Navier-Stokes equation; using boundary conditions of the blood pressures, pericardial fluid, and external loading as the constraints.

Motion of the heart valves is used as a boundary condition in the Navier-Stokes equation in determining the fluid dynamics of blood ejection from the left and right ventricles into the aorta and the lung.

Relationship Between Pressure and Flow in Open Valves

The pressure drop, Δp, across an open heart valve relates to the flow rate, Q, through the valve:

$$a\frac{\partial Q}{\partial t}+bQ^{2}=\Delta p$$

If:

- Inflow energy conserved
- Stagnant region behind leaflets
- Outflow momentum conserved
- Flat velocity profile.

Valves with a Single Degree of Freedom

Usually the aortic and mitral valves are incorporated in valve studies within a single degree of freedom. These relationships are based on the idea of the valve being a structure with a single degree of freedom. These relationships are based upon the Euler equations.

Equations for the aortic valve in this case:

$$\rho\left(\frac{\partial u}{\partial t}+\frac{u\partial u}{\partial x}\right)+\frac{\partial p}{\partial x}=0$$

$$\frac{\partial A}{\partial t}+\frac{\partial}{\partial x}(Au)=0$$

$$A(x,t)=A_0\left(1-[1-\Lambda(t)]\frac{x}{L}\right)^2$$

$$\int_0^L p(x,t)\frac{\partial A}{\partial x}dx=[A_0-A(L,t)]p(L,t)$$

where:

u = axial velocity

p = pressure

A = cross sectional area of valve

L = axial length of valve

Λ (t) = single degree of freedom; when $\Lambda^2(t)=\dfrac{A(L,t)}{A_0}$

Components of Valves

The main parts of the most usual type of valve are the body and the bonnet. These two parts form the casing that holds the fluid going through the valve.

Body

The valve's body is the outer casing of most or all of the valve that contains the internal parts or *trim*. The bonnet is the part of the encasing through which the stem passes and that forms a guide and seal for the stem. The bonnet typically screws into or is bolted to the valve body. Valve bodies are usually metallic or plastic. Brass, bronze, gunmetal, cast iron, steel, alloy steels and stainless steels are very common. Seawater applications, like desalination plants, often use duplex valves, as well as super duplex valves, due to their corrosion resistant properties, particularly against warm seawater. Alloy 20 valves are typically used in sulphuric acid plants, whilst monel valves are used in hydrofluoric acid (HF Acid) plants. Hastelloy valves are often used in high temperature applications, such as nuclear plants, whilst inconel valves are often used in hydrogen applications. Plastic bodies are used for relatively low pressures and temperatures. PVC, PP, PVDF and glass-reinforced nylon are common plastics used for valve bodies.

Bonnet

A bonnet acts as a cover on the valve body. It is commonly semi-permanently screwed into the valve body or bolted onto it. During manufacture of the valve, the internal parts are put into the body and then the bonnet is attached to hold everything together inside. To access internal parts of a valve, a user would take off the bonnet, usually for maintenance. Many valves do not have bonnets; for example, plug valves usually do not have bonnets. Many ball valves do not have bonnets since the valve body is put together in a different style, such as being screwed together at the middle of the valve body.

Ports

Ports are passages that allow fluid to pass through the valve. Ports are obstructed by the valve member or disc to control flow. Valves most commonly have 2 ports, but may have as many as 20. The valve is almost always connected at its ports to pipes or other components. Connection methods include threadings, compression fittings, glue, cement, flanges, or welding.

Handle or Actuator

A handle is used to manually control a valve from outside the valve body. Automatically controlled valves often do not have handles, but some may have a handle (or something similar) anyway to manually override automatic control, such as a stop-check valve. An actuator is a mechanism or device to automatically or remotely control a valve from outside the body. Some valves have neither handle nor actuator because they automatically control themselves from inside; for example, check valves and relief valves may have neither.

Disc

A disc or valve member is a movable obstruction inside the stationary body that adjustably restricts flow through the valve. Although traditionally disc-shaped, discs come in various shapes. Depending on the type of valve, a disc can move linearly inside a valve, or rotate on the stem (as in a butterfly valve), or rotate on a hinge or trunnion (as in a check valve). A *ball* is a round valve member with one or more paths between ports passing through it. By rotating the ball, flow can be directed between different ports. Ball valves use spherical rotors with a cylindrical hole drilled as a fluid passage. Plug valves use cylindrical or conically tapered rotors called plugs. Other round shapes for rotors are possible as well in rotor valves, as long as the rotor can be turned inside the valve body. However not all round or spherical discs are rotors; for example, a ball check valve uses the ball to block reverse flow, but is not a rotor because operating the valve does not involve rotation of the ball.

Seat

The seat is the interior surface of the body which contacts the disc to form a leak-tight seal. In discs that move linearly or swing on a hinge or trunnion, the disc comes into contact with the seat only when the valve is shut. In disks that rotate, the seat is always in contact with the disk, but the area of contact changes as the disc is turned. The seat always remains stationary relative to the body.

Seats are classified by whether they are cut directly into the body, or if they are made of a different material:

- Hard seats are integral to the valve body. Nearly all hard seated metal valves have a small amount of leakage.
- Soft seats are fitted to the valve body and made of softer materials such as PTFE or various elastomers such as NBR, EPDM, or FKM depending on the maximum operating temperature.

Ball Valve

A closed soft seated valve is much less liable to leak when shut while hard seated valves are more durable. Gate, globe, and check valves are usually hard seated while butterfly, ball, plug, and diaphragm valves are usually soft seated.

Stem

The stem transmits motion from the handle or controlling device to the disc. The stem typically passes through the bonnet when present.

In some cases, the stem and the disc can be combined in one piece, or the stem and the handle are combined in one piece.

The motion transmitted by the stem may be a linear force, a rotational torque, or some combination of these (Angle valve using torque reactor pin and Hub Assembly). The valve and stem can be threaded such that the stem can be screwed into or out of the valve by turning it in one direction or the other, thus moving the disc back or forth inside the body. Packing is often used between the stem and the bonnet to maintain a seal. Some valves have no external control and do not need a stem as in most check valves.

Valves whose disc is between the seat and the stem and where the stem moves in a direction into the valve to shut it are normally-seated or front seated. Valves whose seat is between the disc and the stem and where the stem moves in a direction out of the valve to shut it are reverse-seated or back seated. These terms don't apply to valves with no stem or valves using rotors.

Gaskets

Gaskets are the mechanical seals, or packings, used to prevent the leakage of a gas or fluids from valves.

Valve Balls

A valve ball is also used for severe duty, high-pressure, high-tolerance applications. They are typically made of stainless steel, titanium, Stellite, Hastelloy, brass, or nickel. They can also be made of different types of plastic, such as ABS, PVC, PP or PVDF.

Spring

Many valves have a spring for spring-loading, to normally shift the disc into some position by default but allow control to reposition the disc. Relief valves commonly use a spring to keep the valve shut, but allow excessive pressure to force the valve open against the spring-loading. Coil springs are normally used. Typical spring materials include zinc plated steel, stainless steel, and for high temperature applications Inconel X750.

Trim

The internal elements of a valve are collectively referred to as a valve's trim. According to API Standards 600, "Steel Gate Valve-Flanged and Butt-welding Ends, Bolted Bonnets", the trim consists of stem, seating surface in the body, gate seating surface, bushing or a deposited weld for the backseat and stem hole guide, and small internal parts that normally contact the service fluid, excluding the pin that is used to make a stem-to-gate connection.

Valve Operating Positions

Valve positions are operating conditions determined by the position of the disc or rotor in the valve. Some valves are made to be operated in a gradual change between two or more positions. Return valves and non-return valves allow fluid to move in 2 or 1 directions respectively.

Two-Port Valves

Operating positions for 2-port valves can be either shut (closed) so that no flow at all goes through, fully open for maximum flow, or sometimes partially open to any degree in between. Many valves are not designed to precisely control intermediate degree of flow; such valves are considered to be either open or shut. Some valves are specially designed to regulate varying amounts of flow. Such valves have been called by various names such as *regulating*, *throttling*, *metring*, or *needle valves*. For example, needle valves have elongated conically-tapered discs and matching seats for fine flow control. For some valves, there may be a mechanism to indicate by how much the valve is open, but in many cases other indications of flow rate are used, such as separate flow metres.

In plants with remote-controlled process operation, such as oil refineries and petrochemical plants, some 2-way valves can be designated as normally closed (NC) or normally open (NO) during regular operation. Examples of normally-closed valves are *sampling valves*, which are only opened while a sample is taken. Other examples of normally-closed valves are emergency shut-down valves, which are kept open when the system is in operation and will automatically shut by taking away the power supply. This happens when there is a problem with a unit or a section of a fluid system such as a leak in order to isolate the problem from the rest of the system. Examples of normally-open valves are purge-gas supply valves or emergency-relief valves. When there is a problem these valves open (by switching them 'off') causing the unit to be flushed and emptied.

Although many 2-way valves are made in which the flow can go in either direction between the two ports, when a valve is placed into a certain application, flow is often expected to go from one certain port on the upstream side of the valve, to the other port on the downstream side. Pressure regulators are variations of valves in which flow is controlled to produce a certain downstream pressure, if possible. They are often used to control flow of gas from a gas cylinder. A back-pressure regulator is a variation of a valve in which flow is controlled to maintain a certain upstream pressure, if possible.

Shuttle Valve

A shuttle valve is a type of valve which allows fluid to flow through it from one of two sources. Generally a shuttle valve is used in pneumatic systems, although sometimes it will be found in hydraulic systems.

Structure and Function

The basic structure of a shuttle valve is like a tube with three openings; one on each end, and one in the middle. A ball or other blocking valve element moves freely within the tube. When pressure from a fluid is exerted through an opening on one end it pushes the ball towards the opposite end. This prevents the fluid from travelling through that opening, but allows it to flow through the middle opening. In this way two different sources can provide pressure without the threat of backflow from one source to the other.

Applications

A shuttle valve has several applications including:

1. *The use of more switches on one machine:* by using the shuttle valve, more than one switch can be operated on a single machine for safety, and each switch can be placed at any suitable location. This application is normally used with heavy industrial machinery.
2. *Winch brake circuit:* a shuttle valve provides brake control in pneumatic winch applications. When the compressor is operated the shuttle valves direct air to open the brake shoes. When the control valve is centred, the brake cylinder is vented through the shuttle valve, and the brake shoes are allowed to close.
3. *Air pilot control:* converting from air to oil results in locking of the cylinder. Shifting the four-way valve to either extreme position applies the air pilot through the shuttle valve, holding the two air-operated valves open and applying oil under air pressure to the corresponding side of the cylinder. Positioning a manual valve to neutral exhausts the air pilot pressure, closing the two-way valves, and trapping oil on both sides of the cylinder to lock it in position.
4. *Standby and emergency systems:* compressor systems requiring standby or purge gases capability are pressure controlled by

the shuttle valve. This is used for instrumentation, pressure cables, or any system requiring continuous pneumatic input. If the compressor fails, the standby tank—regulated to slightly under the compressor supply—will shift the shuttle valve and take over the function. When the compressor pressure is re-established, the shuttle valve shifts back and seals off the standby system until needed again.

Four-Way Valve

The four-way valve or four-way cock is a fluid control valve whose body has four ports equally spaced round the valve chamber and the plug has two passages to connect adjacent ports. The plug may be cylindrical or tapered, or a ball. It has two flow positions as shown, and usually a central position where all ports are closed. It can be used to isolate and to simultaneously bypass a sampling cylinder installed on a pressurized water line. It is useful to take a fluid sample without affecting the pressure of a hydraulic system and to avoid degassing (no leak, no gas loss or air entry, no external contamination).

It was used to control the flow of steam to the cylinder of early double-acting steam engines, such as those designed by Richard Trevithick. This use of the valve is possibly attributable to Denis Papin.

Because the two "L"-shaped ports in the plug do not interconnect, the four-way valve is sometimes referred to as an "×" port.

Control of Valves: Many valves are controlled manually with a handle attached to the stem. If the handle is turned ninety degrees between operating positions, the valve is called a quarter-turn valve. Butterfly, ball valves, and plug valves are often quarter-turn valves. If the handle is circular with the stem as the axis of rotation in the centre of the circle, then the handle is called a handwheel. Valves can also be controlled by actuators attached to the stem. They can be electromechanical actuators such as an electric motor or solenoid, pneumatic actuators which are controlled by air pressure, or hydraulic actuators which are controlled by the pressure of a liquid such as oil or water. Actuators can be used for the purposes of automatic control such as in washing machine cycles, remote control such as the use of a centralised control room, or because manual control is too difficult such as when the valve is very large. Pneumatic actuators and hydraulic actuators need pressurised air or liquid lines to supply the actuator: an inlet line and an outlet line. Pilot valves are valves which are used to control other valves. Pilot valves in the actuator lines control the supply of air or liquid going to the actuators. The fill valve in a toilet water tank is a liquid level-actuated valve. When a high water level is reached, a mechanism shuts the valve which fills the tank.

In some valve designs, the pressure of the flow fluid itself or pressure difference of the flow fluid between the ports automatically controls flow through the valve.

Other Considerations

Valves are typically rated for maximum temperature and pressure by the manufacturer. The wetted materials in a valve are usually identified also. Some valves rated at very high pressures are available. When a designer, engineer, or user decides to use a valve for an application, he/she should ensure the rated maximum temperature and pressure are never exceeded and that the wetted materials are compatible with the fluid the valve interior is exposed to. In Europe, valve design and pressure ratings are subject to statutory regulation under the Pressure Equipment Directive 97/23/EC (PED).

Some fluid system designs, especially in chemical or power plants, are schematically represented in piping and instrumentation diagrams. In such diagrams, different types of valves are represented by certain symbols.

Valves in good condition should be leak-free. However, valves may eventually wear out from use and develop a leak, either between the inside and outside of the valve or, when the valve is shut to stop flow, between the disc and the seat. A particle trapped between the seat and disc could also cause such leakage.

Endobronchial Valve

An endobronchial valve is an implantable medical device—a small, one-way valve, which is implanted in an airway in the pulmonary system to treat one of several lung conditions. The one-way endobronchial valve is typically implanted such that when a patient exhales, air is able to flow through the valve and out of the lung compartment that is fed by that airway, but when the patient inhales, the valve closes and blocks air from entering that lung compartment. Thus, an implanted endobronchial valve typically helps a lung compartment to empty itself of air. This has been shown to be beneficial in the treatment of emphysema, where lungs lose their elasticity and thus cannot contract sufficiently to exhale air, leading to air trapping and hyperaeration. When one or more diseased portions of an emphysematous lung are made to deflate and collapse, other healthier portions of the lung have more room in the chest cavity to inhale and exhale, pressure is removed from the diaphragm, and even the heart may function better as the hyperinflated lung becomes smaller. Endobronchial valves have also been shown to be beneficial in treatment of persistent air leaks in the lungs. It has also been theorized that endobronchial valves may be useful for treatment of tuberculosis and its complications.

Endobronchial valves were developed primarily by the start-up medical device company Emphasys Medical (now Pulmonx - Redwood City, California) as a minimally invasive alternative to lung volume reduction surgery for emphysema. In lung volume reduction surgery, a surgeon cuts into a patient's chest and cuts out one or more most diseased portion(s) of a lung. Endobronchial valves were designed to replicate the effects of that procedure without requiring incisions, by simply allowing the most diseased portions of the lung to collapse. Emphasys was purchased by Pulmonx in 2009, and Pulmonx currently markets the Zephyr® endobronchial valve (developed by Emphasys) in Europe, Australia, China and many other locations outside the U.S. Pulmonx also sells the Chartis® Pulmonary Assessment System, which is a patient assessment tool used with endobronchial valves to help physicians target appropriate lung compartments for treatment. Another company, Spiration (Seattle, Washington), developed a different type of endobronchial valve and was acquired by Olympus in 2010.

Endobronchial valves may be the first successful medical device treatment of emphysema, a disease that affects billions of patients worldwide and has no known cure. Endobronchial valves are typically implanted using a flexible delivery catheter advanced through a bronchoscope, and thus they are very minimally invasive. The valves are also removable if they are not working properly. When properly used and correctly implanted, endobronchial valves have been shown to improve patients' ability to breathe, walk, exercise and perform their daily activities. Although not yet approved by the Food and Drug Administration for emphysema treatment in the United States, endobronchial valves are available in many other locations throughout the world and should be available in the U.S. eventually.

Directional Control Valve

Directional control valves are one of the most fundamental parts in hydraulic machinery. They allow fluid flow into different paths from one or more sources. They usually consist of a spool inside a cylinder which is mechanically or electrically controlled. The movement of the spool restricts or permits the flow, thus it controls the fluid flow.

Nomenclature: The spool (sliding type) consists of lands and grooves. The lands block oil flow through the valve body. The grooves allow oil to flow around the spool and through the valve body. There are two fundamental positions of directional control valve namely normal position where valve returns on removal of actuating force and other is working position which is position of a valve when actuating force is applied.

Classification

Directional control valves can be classified according to:

- number of ports
- number of positions
- actuating methods
- type of spool.

Example: A 5/2 directional control valve would have five ports and two spool positions.

Number of Ports: According to total number of entries or exits connected to the valve through which fluid can enter the valve or leave the valve there are types like two way, three way, four way valves.

Number of Positions: Including the normal and working positions which a valve spool can take there are types like two position, three position valves.

Actuating Methods

Manually Operated: Manually operated valves work with simple levers or paddles where the operator applies force to operate the valve. Spring force is used to recover the position of valve.

Mechanically Operated: Mechanically operated valve apply forces by using cams, wheels, rollers etc. Hence these valves are subjected to wear.

Hydraulically Operated: An hydraulically operated DCV works at much higher pressures than its pneumatic equivalent, they must therefore be far more robust in nature so are precision machined from higher quality and strength materials.

Solenoid Operated

They have a Vast Use in hydraulic industry. These valves make use of electromechanical Solenoid for sliding of the spool. Electrical power is readily available and these valves are extensively used but electrical solenoid cannot generate large forces unless supplied with larger electrical energy which makes there use limited to low actuating forces. Heat generation poses the threat to use of these valves when held in one position longer. In practice combination of these actuators. Such as low power solenoid valve is used to operate the other hydraulic valve. Here solenoid valve starts the flow of fluid which in turn operates hydraulic valve.

Type of Spool

Spool is of two types namely sliding and rotary. Sliding spool is cylindrical in cross section lands and grooves are also cylindrical whereas rotary valve has sphere like lands and grooves in the form of holes drilled through sphere.

Specification

They are generally specified using the number of ports and the number of switching positions. It can be represented in general form as n_p/n_s, where n_p is the number of ports connected to the direction control valve and n_s the number of switching positions.

In addition, the method of actuation and the return method can also be specified. A hypothetical valve could be specified as 4-way, 3-position direction control valve or 4/3 DCV since there are four ports and three switching positions for the valve. In this example, one port is called the pressure port which is connected to the pump; one port is the tank port and is connected to the tank (or reservoir); and the two remaining ports are called working ports and are connected to the actuator. Apart from characteristics of valve the fluid suitable for valve, working temperature and viscosity also thought upon before selecting a particular type of valve.

Symbolic Representation

while working with layouts of hydraulic machinery it is cumbersome to draw actual picture of every valve and other components. Instead of pictures symbols are used for variety of components in the hydraulic system to highlight the functional aspects. Symbol for directional control valve is made of number of square boxes adjacent to each other depending on the number of positions. Connections to the valve are shown on these squares by capital letters. Usually they are named only in their normal position and not repeated in other positions. Actuation system of the valve is also designated in its symbol.

Two Way Two Position Directional Control Valve

Gate valve is example of 2W/2P directional control valve which either turns on or off the flow in normal or working position depending on need of application. Here arrow indicates that fluid flow is taking place whereas other position shows cut-off position.

Four Way Two Position Directional Control Valve

4 way 2 position valve has four connections to it and two valve positions one of them is normally open.

Control Valves

Control valves are valves used to control conditions such as flow, pressure, temperature, and liquid level by fully or partially opening or closing in response to signals received from controllers that compare a "setpoint" to a "process variable" whose value is provided by sensors that monitor changes in such conditions.

The opening or closing of control valves is usually done automatically by electrical, hydraulic or pneumatic actuators. Positioners are used to control the opening or closing of the actuator based on electric, or pneumatic signals. These control signals, traditionally based on 3-15psi (0.2-1.0bar), more common now are 4-20mA signals for industry, 0-10V for HVAC systems, and the introduction of "Smart" systems, HART, Fieldbus Foundation, and Profibus being the more common protocols.

Types of Control Valve Bodies

The most common and versatile types of control valves are sliding-stem globe and angle valves. Their popularity derives from rugged construction and the many options available that make them suitable for a variety of process applications, including severe service. Control valve bodies may be categorized as below:

- Angle valves
  - Cage-style valve bodies
  - DiskStack style valve bodies
- Angle seat piston valves
- Globe valves
  - Single-port valve bodies
  - Balanced-plug cage-style valve bodies
  - High capacity, cage-guided valve bodies

  - Port-guided single-port valve bodies
  - Double-ported valve bodies
  - Three-way valve bodies
- Diaphragm Valves
- Rotary valves
  - Butterfly valve bodies
  - V-notch ball control valve bodies
  - Eccentric-disk control valve bodies
  - Eccentric-plug control valve bodies
- sliding cylinder valves
  - Directional control valve
  - spool valve
  - piston valve
- air operated valves
  - air operated valve
  - relay valve
  - air operated pinch valve

Types of Control Valve Trims

- Window type cage
- Drilled Hole Cage
- Mutli Turn Tortures path Trim

Selection Parameters

When selecting a valve for a specific service, several parameters of valve design and performance must be evaluated. This discussion will focus on the more important specifications and attributes to be considered when selecting a control valve. They a pressure / temperature ratings, material compatibility, capacity, flow characteristics, shutoff, cavitation and flashing, noise, end connections, economy, and maintenance.

Pressure Ratings

When selecting a valve for a given service we must consider both the inlet pressure rating and the maximum pressure drop rating of the control valve. Inlet pressure ratings refer to the capabilities of the pressure retaining components such as bodies and bonnets. This capability is commonly expressed in terms of ANSI Class ratings.

ANSI stands for the American National Standards Institute. ANSI Class ratings of 150, 300, 600, 900, 1500, or 2500 pounds are commonly used. Generally speaking, valves with higher ANSI Class ratings have heavier or thicker pressure retaining components.

ANSI ratings define requirements for classes over 2500 and other special categories. While ANSI standards are commonly referenced, other standards such those published by the API (American Petroleum Institute) are also used to define pressure/temperature capabilities. A valve with a given ANSI Class rating is likely to be capable of withstanding inlet pressures much greater than the nominal ANSI class. For example, an ANSI Class rating of 150 does not limit the valve to 150 psig. In fact, ANSI Class 150 carbon steel valves bodies are typically rated at 285 psig at lower temperatures. Within the same ANSI Class, different materials have different pressure-temperature capabilities. Therefore, when a valve is selected that does not quite meet the service pressure requirements, it may be possible to select the same valve in the same ANSI class but in a different material that will meet pressure/temperature requirements. In, many instances, this can be more economical than selecting a heavier body with a higher ANSI Class rating. Increased temperatures limit the maximum inlet pressure rating, the higher the temperature, the lower the maximum inlet pressure rating.

Pessure Drop Ratings

Sometimes more limiting than the ANSI Class rating is the pressure drop (ΔP) rating.

Pressure drop is defined as the difference between the inlet and the outlet pressures of the valve. While control valve inlet pressure ratings just discussed are determined by the strength of pressure retaining parts such as bodies and bonnets, control valve pressure drop capabilities are more a function of the valve's internal parts; closure members, stems, shafts, bearings, and so forth. Flowing pressure drop refers to the difference between downstream and upstream pressure, while the valve is throttling. Flowing pressure drops are highest when the valve closure member is positioned near its seat. In this position, downstream pressure is greatly reduced, velocity in the valve is high and the forces of fluid flow acting on the closure member (plug, ball, or disk) are increased. As a general rule of thumb, shutoff pressure drop is equivalent to upstream pressure. When the valve is closed, downstream pressure is essentially zero so the full inlet pressure is the pressure drop. Exceptions to this guideline are when downstream back-pressure acts to reduce pressure drop or, where a downstream vacuum increases pressure drop. Pressure containment is sometimes accompanied by a change in energy states; a change from static pressure - or potential energy - to a form of kinetic energy such as heat, vibration, or noise. Any of these forces can produce damaging effects oil control valve components. Depending on how the forces associated with pressure drop are applied to closure members and controlling components, either flowing or shutoff pressure drop may become a limiting factor in control valve selection. In rotary shaft valves, fluid forces acting on the closure member may stress or destabilise the member itself.

In addition, these forces are ultimately converted to torque that applies stress to shafts, bearings, pins, and other components connected to the closure member or shaft. In globe or sliding stem valves, forces associated with pressure drop result in compressive and tensile forces on the stem, stress on the stem-to-plug connection, and instability of the valve plug. When applying a control valve, it must be determined that the valve selected has maximum flowing and shutoff pressure drop ratings that match the service conditions. Control valve manufacturers determine and publish both flowing and shutoff pressure drop capabilities. The control valve actuator must be sized to effectively oppose the forces of flow o the valve closure member. Therefore, unbalanced globe valves and many rotary-shaft valves may require extreme amounts of actuator force under certain conditions.

Temperature Ratings

Temperature capabilities are essentially a function of the materials used. However, as we saw in the ANSI pressure-temperature ratings charts, temperature a pressure ratings are interactive; that is, the higher the temperature, the lower the pressure capability. Each component of the valve has a temperature rating; the body, trim, packing, seals, gaskets, and so forth. Therefore, the temperature capability of the valve assembly is defined by the component with the most restrictive rating. In the workbook example, the bonnet bolts limit the maximum service temperature to 450 degrees. Since all other materials are capable of higher temperatures, the bolting specification is the limiting factor. Most manufacturers have checking systems in place to prevent this inconsistency in material selection. In some instances, the effects of thermal expansion and contraction may be more limiting than the nominal temperature rating of the construction material used.

This is especially true when large or long valve components are used in extremely high or low temperature service. In such instances, the valve assembly may have a more limited temperature capability than the materials from which it was manufactured. Manufacturers generally consider these factors and rate their products accordingly.

Material Selection

Material specification is an important step in control valve selection. Material selection has great impact on the suitability of a valve for a given medium, the life of the valve, and the cost of the valve. Factor's which govern material selection are:

Chemical compatibility with a given process fluid

Physical ability to withstand the affects or wear, pressure drop, and erosive fluids Trim materials are often made of more durable materials than the body, because they are in the flowstream and subject to more wear.

Heat treated or hard-faced stainless steels are commonly used for valve trim. Carbon steel valve bodies are popular because of their economy and broad range of availability. Stainless steel valve bodies may be specified when exceptionally high or low temperature capabilities are required or for resistance to some corrosive process fluids. Soft valve components such as seats, seals, and packing are available in a broad range of natural and synthetic materials. Application requirements such as temperature and chemical compatibility dictate the materials used. In erosive applications, trim parts can be hard-faced or manufactured from very durable materials such as Alloy 6, tungsten carbide, ceramics, and other new speciality materials. For extremely corrosive applications, bodies and trim may be made from one of a broad range of nickel based alloys such as the Hastelloys, Inconels, and Monels. As a word of caution, many different standards exist for describing materials. Materials can be described with common names such as "stainless steel", or with proprietary names, such as Inconnel, which is a trademark of the International Nickel Company. These rather general terms could apply to any one of a number of specific materials within that family. For instance, a general reference to Inconel could be interpreted as Inconel 718 or Inconel 600. The 718 material has a yield strength of 180,000 psi while the 600 material has a yield strength of only 25,000 PSI. Therefore, with this wide range of capability in the same family of material, it is important to specify the exact material desired.

Capacity

Capacity is defined as the rate of flow through a valve under stated conditions as, expressed as a "flow coefficient". The flow coefficient is supplied by the valve manufacturer and is useful for determining the size of a given control valve needed for a given flow requirement. Different flow coefficients are used for each of the physical states - liquid, gas, and steam. Liquid capacity is expressed as the coefficient Cv. Cv is defined as the number of gallons of water at 60 degrees F. that will flow through the valve in one minute at a pressure drop of 1 psi at stated conditions.

The stated conditions will include percent of rated travel. Gas and steam capacities are expressed with different flow coefficients because they behave differently than liquids. In arriving at these coefficients, allowances are made for the effects of compressibility and other important factors. The coefficients Cg, - gases, and Cs - for steam, are commonly used. Specific ANSI standards define commonly accepted procedures for determining flow coefficients. These standards define procedures for determining flow coefficients by actual pipeline tests and by calculation.

The procedures used by different manufacturers to determine flow coefficients can vary so it is a good practice to know if the manufacturer typically rates their valves conservatively or optimistically. Tested results may be preferred as they may be more accurate. Rotary-shaft valves, with the exception of the eccentric rotary-plug design, provide the highest liquid flow coefficients. This is expected as these valves have a straight through flow path with minimal obstruction. While it might be expected that the full ball valve would have the highest flow coefficient, recall that the full ball valve discussed earlier had a reduced port to provide accurate throttling control. Because, globe valves have a more tortuous flow path than rotary-shaft valves, we would expect them to have somewhat diminished capacity compared to rotary valves. When flowing gasses, the difference in capacity between the valve families is less pronounced.

This is because liquid flow coefficients are dependant on how tortuous the flow path is, while gas flow coefficients are more a function of the minimum cross-sectional flow area of the control valve. Closely related to capacity, recovery is a relative term used to describe how downstream pressure is affected by the design and geometry of the control valve. For example, globe valves tend to be low recovery valves. That is, they have a relatively tortuous flow path, which limits capacity. On the other hand, most rotary (ball and butterfly) valves have a "line of sight" flow path, which provides high capacity so they are referred to as "high recovery" valves. Though high recovery valves have the benefit of being high capacity valves, they are also more susceptible to troublesome and damaging phenomena such as cavitation.

Flow Characteristics

Flow characteristics determine how the flow coefficient changes in response change in the valve position; in other words, how the rate of flow will be changed when the valve receives a signal from the sensing instrument. Selection of the best flow characteristic for a process is a relatively involved process, so for the purposes of this discussion, we will simply identify the more common characteristics. Quick Opening - An inherent flow characteristic in which there is maximum flow coefficient with minimum travel. Linear - A inherent characteristic which can be represented ideally by a straight line plot of flow coefficient versus percent of rated travel. (Equal increments of travel produce equal increments of flow at a constant pressure drop) Equal percentage - An inherent flow characteristic in which a given percentage change of valve travel will produce and equal percentage change in the exiting flow coefficient. Flow characteristics of cage guided valves can be determined by the shape of the cage windows. Stem or post guided valves can be characterised by the shape of the plug, and most rotary valves have a flow characteristic inherent to their basic design. The flow characteristics discussed are "inherent" characteristics. That is, they describe the flow characteristics under static conditions of constant pressure and pressure drop. In actual service conditions, changes in pressure and pressure drop will change the flow characteristic dramatically. Therefore, a distinction is made between "inherent" and "installed" characteristics. The installed characteristic takes into account changes in pressure drop and other conditions in the system.

Shutoff Ansi Class Seat Leakage

Shutoff is ordinarily stated in terms of classes of seat leakage defined in the American National Standard for Control Valve Seat Leakage. In actual service, shutoff leakage depends on many factors including pressure drop, temperature, the condition of the sealing surfaces, and the force load on the seat -, which is a function of actuator force available. Since shutoff ratings are based on standard test conditions, which may be very different from actual service conditions, service leakage cannot be absolutely predicted. However, the ANSI shutoff classes provide a good basis for comparison among valves of similar configuration. ANSI Classes Compared As we identify the different seat leakage standards, we can roughly calculate the seat leakage of a typical 3-inch globe body that would conform to each of the leak classes.

First, because ANSI Class two, three, and four leakage is expressed as a percentage of rated capacity, we'll have to calculate the normal wide open flow of our three inch valve under test conditions. The basic formula for flow is C, times the square-root of AP so we'll have to know the C, of the valve and the pressure drop of our setup. The maximum rated C, of 140 comes from the manufacturers literature and the pressure drop of 50 psi is one of the test conditions in the ANSI Standard. Solving the equation, we find that our 3-inch valve will produce a maximum flow of approximately 1,000 gallons per minute under test conditions.

Class I shutoff does not require testing but is mutually defined by and agreed to by the user and supplier. It is a special classification that might apply to a valve with a higher leakage class rating which has been modified for some purpose. Class II shutoff allows leakage of up to one-half of one percent (0.5%) of the rated capacity of the valve using air or water as the test medium at a pressure drop of 50 psid. In our example, leakage of the specified valve is five gallons per minute. Class III shutoff allows leakage of up to one-tenth of one percent (0.1 %) of the rated capacity of the valve again using air or water as the test medium with a 50 psid pressure drop. This is one-fifth the leakage of Class II and as an the example, is one gallon per minute. Class IV shutoff allows leakage of one one-hundredth of one percent (.01 %) of the rated capacity of the valve under the same test conditions as above.

Leakage is slightly less than one pint per minute. Class V standards become more stringent and allow only 0.0005 millilitres of water per minute per inch of port diametre at a minimum test pressure drop of 100 psid. In the example, wide open flow should be increased to about 1400 gallons per minute because of the increased test pressure. However, because of the more demanding requirements, an eyedropper could be used to measure leakage accumulated in one minute. Class VI standards are very demanding. Instead of water as a test medium, air or nitrogen is used with a pressure drop of 50 psid. The allowable leakage for different nominal port diametres is expressed in both millilitres per minute and bubbles per minute. Allowable leakage does not follow a linear scale but is identified for port diametres through 8-inches. Obviously, this leak class provides very tight shutoff.

In actual application, not all throttling valves need to provide tight shutoff. Block valves placed around the control valve provide the tight shutoff function. Over specifying control valve shutoff has been identified as one of the greatest unnecessary costs incurred during control valve selection. What typically happens is that every specifier in the chain - the designer, plant manager, engineer, and so on -adds a safety margin to the specifications; each person who reviews the plans and specifications increases the shutoff requirement. The result is good - though not always necessary - shutoff, at a progressive penalty in cost. In some instances, tight shutoff should be specified even though it not a specific process requirement. For instance, when flowing toxic or flammable fluid, tight shutoff is often specified for safety reasons. In severe services involving erosive fluids, high pressure drops, or cavitation (discussed later), tight shutoff may be specified to reduce wear or erosion of closure members and seats.

Noise Considerations

Aerodynamic noise is generated by the turbulence associated with control of gas, steam, or vapours. While generally thought of as accompanying high capacity, high pressure systems, damaging noise levels can be produced in a two-inch line with as little as a 200 psi pressure drop. Concern for high levels of noise is a twofold issue. First, noise control is an environmental issue as OSHA (the Office for Safety and Health Administration) has determined safe and acceptable limits for control valve noise as it affects the working environment. Secondly, research indicates that noise levels above the recommended limit of approximately 110 decibels can result in mechanical damage to control valves, their associated parts and piping. During the valve sizing process, noise can be predicted by using well established mathematical techniques.

When unacceptable noise levels are predicted, noise abatement equipment and techniques are available to reduce noise to acceptable levels. Noise control can be provided by source treatment at the origin (generally the control valve), by path treatment downstream of the origin, or by both. Source treatment strategies generally involve special valve trims or accessories designed to reduce turbulence of the process fluid within the valve. Source treatments actually limit the amount of noise generated by a control valve. For cage guided valves, special cage designs are used. For other types of valves, diffusing components are inserted downstream to reduce noise generation within the valve. Path treatments such as pipeline insulation or the use of heavy wall pipe do not eliminate the noise but rather reduce the intensity, of the sound transmitted into the working environment. Silencers, on the other hand, do absorb part of the noise energy and therefore reduce the sound intensity in both the flowstream and the working environment.

End Connections

End connection selection is generally a simple question of whether the desired end connection style is available for the type of valve being considered. In some instances, however, end connection style is dictated by safety standards or plant policies.

Screwed connections can be used in sizes up to 2-inch in general service applications. Flanged connections are available in several styles. Flat face flanges provide full face contact. They are commonly used in low pressure service with cast iron or brass valves. Raised face flanges are the most popular of all end connections. They have a grooved gasketing surface to provide a tight gasket seal and to help prevent gasket blowout.

These flanges can be used with temperatures through 1500*F and pressures to 6000 psig. The ring type joint flange is similar to the raised face flange except that a compressible metal ring is wedged into two mating grooves when the flanges are tightened. These are used for high pressure service to 15,000 psi but are not used for high temperature service. Separable flanges are manufactured independently of the valve body and are secured to the body with split rings.

They are economically beneficial when bodies must be made of costly alloys for corrosion, and erosive services. Because the process fluid does not normally contact the flanges, flanges can be manufactured from materials that are more economical. Welded end connections have the advantage of being leak tight at all pressures and temperatures. Obviously, they are only used with weldable materials and make it difficult to remove the valve from the pipeline Socket weld ends are dimensionally the same regardless of pipe schedule and are generally used in sizes through two inches. Butt weld ends are prepared by beveling, each end of the valve to match a similar bevel on the pipe. End preparation is different for different schedules of pipe. This end connection is typically used with valve sizes over two-inches.

Valve Sizing

The choice of valve for a particular application depends on various factors such as the fluid flow rate, the process temperatures and pressures, and whether the fluid is corrosive or abrasive. An important design consideration for safety is the maximum pressure and temperature of the flowing fluid as this will dictate a safe thickness for the body material and may determine which materials can be used. The capacity of a valve is measured in Cv, Kv or Av and can be determined experimentally for a valve by testing it, for example by measuring the amount of water that flows through it at a certain pressure differential across the valve.

The valve needed for an application will depend on whether the fluid is a liquid, gas or multi-phase fluid and on the properties of the fluid. Normally a number of process conditions - each of which will specify the inlet and outlet pressure, temperature, flow rate and maybe other fluid details - are used to calculate a required capacity, and to estimate the noise and the cavitation. Valves often have interchangeable internal components, called the trim, which can be selected to modify the valve to accommodate particular process conditions. For example perforated trim can be used to reduce the noise level by splitting the flow into multiple independent streams.

Due to the complexity of the fluid flow, noise and cavitation calculations often required to size and select an appropriate control valve many manufacturers provide a computer program to assist their customers. For example Severn Glocon and Dresser provide downloadable sizing programs. A small number of software companies have specialist products for this market such as Sahir Projects, Conval, Millstream Software Valve Sizing Program and Configurator solutions Configurator Solutions.

Variable Valve Timing

In internal combustion engines, variable valve timing (VVT), also known as Variable valve actuation (VVA), is a generalised term used to describe any mechanism or method that can alter the shape or timing of a valve lift event within an internal combustion engine. VVT allows the *lift*, *duration* or *timing* (in various combinations) of the intake and/or exhaust valves to be changed while the engine is in operation. Two-stroke engines use a power valve system to get similar results to VVT. There are many ways in which this can be achieved, ranging from mechanical devices to electro-hydraulic and camless systems.

The valves within an internal combustion engine are used to control the flow of the intake and exhaust gases into and out of the combustion chamber. The timing, duration and lift of these valve events has a significant impact on engine performance. In a standard engine, the valve events are fixed, so performance at different loads and speeds is always a compromise between driveability (power and torque), fuel economy and emissions. An engine equipped with a variable valve actuation system is freed from this constraint, allowing performance to be improved over the engine operating range.

Strictly speaking, the history of the search for a method of variable valve opening duration goes back to the age of steam engines when the valve opening duration was referred to as "steam cut-off". Almost all steam engines had some form of variable cut-off. That they are not in wide use is a reflection that they are all lacking in some aspect of variable valve actuation.

The desirability of being able to vary the valve opening duration to match an engine's rotational speed first became apparent in the 1920s when maximum allowable RPM limits were generally starting to rise. Until about this time an engine's idle RPM and its operating RPM were very similar, meaning that there was little need for variable valve duration.

It was in the 1920s that the first patents for variable duration valve opening started appearing – for example United States patent U.S. Patent 1,527,456. A surprising fact is that from these first patents until the appearance of the helical camshaft there has never been a really practical and useful variable duration camshaft.

Overview

Piston engines normally use poppet valves for intake and exhaust. These are driven (directly or indirectly) by cams on a camshaft. The cams open the valves (lift) for a certain amount of time (duration) during each intake and exhaust cycle. The timing of the valve opening and closing is also important. The camshaft is driven by the crankshaft through timing belts, gears or chains.

The profile, or position and shape of the cam lobes on the shaft, is optimized for a certain engine revolutions per minute (RPM), and this tradeoff normally limits low-end torque, or high-end power. VVT allows the cam timing to change, which results in greater efficiency and power, over a wider range of engine RPMs. An engine requires large amounts of air when operating at high speeds. However, the intake valves may close before enough air has entered each combustion chamber, reducing performance. On the other hand, if the camshaft keeps the valves open for longer periods of time, as with a racing cam, problems start to occur at the lower engine speeds. This will cause unburnt fuel to exit the engine since the valves are still open. This leads to lower engine performance and increased emissions. For this reason, pure racing engines which are designed to idle at speeds close to 2,000 rpm, cannot idle well at the lower speeds (around 800 rpm) expected of a road car. Pressure to meet environmental goals and fuel efficiency standards is forcing car manufacturers to use VVT as a solution. Most simple VVT systems advance or retard the timing of the intake or exhaust valves. Others (like Honda's VTEC) switch between two sets of cam lobes at a certain engine RPM. Furthermore Honda's i-VTEC can alter intake valve timing continuously.

History

Steam Engines

The first variable valve timing systems came into existence in the nineteenth century on steam engines. Stephenson valve gear, as used on early steam locomotives, supported variable cutoff, that is, changes to the time at which the admission of steam to the cylinders is cut off during the power stroke. Early approaches to variable cutoff coupled variations in admission cutoff with variations in exhaust cutoff. Admission and exhaust cutoff were decoupled with the development of the Corliss valve. These were widely used in constant speed variable load stationary engines, with admission cutoff, and therefore torque, mechanically controlled by a centrifugal governor and trip valves. As poppet valves came into use, simplified valve gear using a camshaft came into use. With such engines, variable cutoff could be achieved with variable profile cams that were shifted along the camshaft by the governor. This is now coming in system.

Aircraft

Some versions of the Bristol Jupiter radial engine of the early 1920s incorporated variable valve timing gear, mainly to vary the inlet valve timing in connection with higher compression ratios. The Lycoming R-7755 engine had a Variable Valve Timing system consisting of two cams that can be selected by the pilot. One for take off, pursuit and escape, the other for economical cruising.

Automotive Use

In 1958 Porsche made application for a German Patent, also applied for and published as British Patent GB861369 in 1959. The

Porsche patent used an oscillating cam driven via a push/pull rod from an eccentric shaft or swashplate. The cam was Desmodromic having opening and closing cam surfaces which operated the valve by a bifurcated rocker and ball joint. Being Desmodromic meant there was no valve spring. The cam pivot was adjustable for height, as the push/pull rod length was constant this rotated the cam so the lift and duration increased. A compensating link moved the rocker pivot to match the cam's position. The adjustment of the cam pivot could be by mechanical linkage to a screw thread, hydraulic from engine driven pump with spill valve or from a engine speed governor. At present it is unknown if any working prototype was ever made.

Fiat was the first auto manufacturer to patent a functional automotive variable valve timing system which included variable lift. Developed by Giovanni Torazza in the late 1960s, the system used hydraulic pressure to vary the fulcrum of the cam followers (US Patent 3,641,988). The hydraulic pressure changed according to engine speed and intake pressure. The typical opening variation was 37%.

In September 1975, General Motors (GM) patented a system intended to vary valve lift. GM was interested in throttling the intake valves in order to reduce emissions. This was done by minimising the amount of lift at low load to keep the intake velocity higher, thereby atomizing the intake charge. GM encountered problems running at very low lift, and abandoned the project.

Alfa Romeo was the first manufacturer to use a variable valve timing system in production cars (US Patent 4,231,330). The 1980 Alfa Romeo Spider 2.0 L had a mechanical VVT system in SPICA fuel-injected cars sold in the United States. Later this was also used in the 1983 Alfetta 2.0 Quadrifoglio Oro models as well as other cars. The system was engineered by Ing Giampaolo Garcea in the 1970s.

Honda's REV motorcycle engine employed on the Japanese market-only Honda CBR400F in 1983 provided a technology base for VTEC.

In 1986, Nissan developed their own form of VVT with the VG30DE(TT) engine for their MID4 Concept. Nissan chose to focus their NVCS (Nissan Valve-Timing Control System) mainly on torque production at low to medium engine speeds, because, the vast majority of the time, automobile engines will not be operated at extremely high speeds. The NVCS system can produce a smooth idle and high amounts of torque at low to medium engine speeds.

The VG30DE engine was first used in the 300ZX (Z31) 300ZR model in 1987. It was the first production car to use electronically controlled VVT technology. In 1987 Nissan also sold the Gloria, Leopard, and Cedric, all of which could come powered by the VG20DET engine which also utilised Nissans NVCS valve timing system.

The next step was taken in 1989 by Honda with the VTEC system. Honda had started production of a system that gives an engine the ability to operate on two completely different cam profiles, eliminating a major compromise in engine design. One profile designed to operate the valves at low engine speeds provides good road manners, low fuel consumption and low emissions output. The second is a high lift, long duration profile and comes into operation at high engine speeds to provide an increase in power output.

The VTEC system was also further developed to provide other functions in engines designed primarily for low fuel consumption. The first VTEC engine Honda produced was the B16A which was installed in the Integra, CRX, and Civic hatchback available in Japan and Europe. In 1991 the Acura NSX powered by the C30A became the first VTEC equipped vehicle available in the US. VTEC can be considered the first "cam switching" system and is also one of only a few currently in production.

In 1991, Clemson University researchers patented the Clemson Camshaft which was designed to provide continuously variable valve timing independently for both the intake and exhaust valves on a single camshaft assembly. This ability makes it suitable for both pushrod and overhead cam engine applications.

In 1992, Porsche introduced VarioCam its 968 model which provided continuously variable valve timing for the intake valves.

In 1992, BMW introduced the VANOS system. Like the Nissan NVCS system it could provide timing variation for the intake cam in steps (or phases), the VANOS system differed in that it could provide one additional step for a total of three. Then in 1996 the Double Vanos system was introduced which significantly enhances emission management, increases output and torque, and offers better idling quality and fuel economy. Double Vanos was the first system which could provide electronically controlled, continuous timing variation for both the intake and exhaust valves.

Ford began using Variable Cam Timing in 1998 for the Ford Sigma engine and the Ford Zetec engine.

In 1999, Porsche introduced VarioCam Plus on its 911 Turbo which combined continuous valve timing and two stage valve lift on the intake valves.

In 2001, BMW introduced the Valvetronic system. The Valvetronic system can continuously and precisely vary intake valve lift, and in addition, the independent Double VANOS system can concurrently vary the timing for both the intake and exhaust valves. The precise control the system has over the intake valves allows for the intake charge to be controlled entirely by the intake valves, eliminating the need for a throttle valve and greatly reducing pumping loss. The reduction of pumping loss accounts for a 10-15% increase in power output and fuel economy.

Ford became the first manufacturer to use variable valve timing in a pickup-truck, with the top-selling Ford F-series in the 2004 model year. The engine used was the 5.4 L 3-valve Triton.

In 2005, General Motors offered the first Variable Valve timing system for pushrod V6 engines, *LZE* and *LZ4*.

In 2007, Daimler Chrysler became the first manufacturer to produce a cam-in-block engine with independent control of exhaust cam timing relative to the intake. The 2008 Dodge Viper uses Mechadyne's concentric camshaft assembly to help boost power output to 600 bhp (450 kW).

In 2009, Fiat Powertrain Technologies introduced the Multiair system in Geneva Motor Show. The Multiair is a hydraulically actuated variable valve timing system, which gives full control over valve lift and timing. The new technology is available in Alfa Romeo MiTo starting from September 2009.

In 2009, Porsche introduced an enhanced version of VarioCam Plus on its 911 GT3 including the previous variable valve timing and two stage valve lift on the intake valves but with additional variable timing of the exhaust valves.

Diesel Engines

In 2010, Mitsubishi developed and started mass production of its 4N13 1.8 L DOHC I4 world's first passenger car diesel engine that features a variable valve timing system.

Types

Purely Mechanical: Note that the types briefly described below relate to the ability to vary primarily the valve opening duration in a continuous fashion by mechanical means. To be able to alter the valve opening duration in a continuous fashion by mechanical means has proven to be a surprisingly difficult task. To achieve this aim has been a "holy grail" for many years for inventors and car company research departments alike. There have only ever been two systems of this type that have been successfully used in cars sold for normal road use – BMW's Valvetronic (and some others operating on this principle, but BMW was the first) and the Rover VVC system. Essentially there are only a very limited number of basic classes (and subclasses) of mechanism known at present that can be employed (in a reasonably practical sense) to vary the valve opening duration of an engine. There are, of course, other proposed schemes and patents for realising continuous variable valve opening duration by mechanical methods. Generally speaking none of these other methods are usable in any practical sense. Those described below are essentially the only workable types.

Oscillating Cam. BMW's Valvetronic, the Nissan VVEL and Toyota Valvematic all belong to the "oscillating cam" class of mechanism. This is a class that dates back to the days of steam engines where the amount of steam entering the cylinder was regulated by the steam "cut-off" point (although the valve gear of most steam engines was not of this type). The existing systems vary widely in their appearance but all have the same underlying principle. This involves a conventional cam lobe (Valvetronic and Valvematic) or eccentric and connecting rod (VVEL) to generate an oscillating or rocking motion in a part cam lobe which then acts on a follower and the valve. The part cam lobe consists of a section of base circle and a section of lobe flank. The position of the part lobe can be adjusted so that the follower is exposed to varying amounts of base circle or lobe flank.

At one extreme is all base circle resulting in zero lift and zero duration, at the other extreme is all flank giving maximum lift and maximum duration. There is a continuous range of lift/duration settings between the two limits but the duration and lift are always directly proportional to each other. The Valvetronic has been widely used by BMW and has proved to be very reliable in service. However it has a major flaw in that the amount of duration is linked in direct proportion to the amount of lift. This has the effect of lessening the effectiveness of the potentially very wide duration range. The Valvetronic has only ever been used on the intake valves. A main strength of Valvetronic – the ability to produce valve openings of very short duration and very low (down to zero) lifts has little useful relevance for the exhaust valve. Because of the linked duration/lift situation the Valvetronic (and the other similar systems) act essentially in their behaviour as a variable lift systems rather than genuinely variable duration arrangement. Despite BMW's claims, the Valvetronic has only a relatively minor beneficial effect on engine power and fuel consumption – something of the order of 10%. More recent developments of the basic oscillating cam principle have been used in the Nissan VVEL and the Toyota Valvematic. These are neater and more compact in size than the Valvetronic but still have similar limitations in their performance. Mention should be made of the Honda AVTEC system. This system is notable and interesting in that, despite its appearance, it is variable lift only - the duration remains fixed. This fact is made quite clear in the text of the Honda patent.

Eccentric Cam Drive. There is little question that this is an extremely ingenious system. Despite its cleverness it has serious basic design problems in practice. Only the Rover company ever used this system. The system operates through an eccentric disc mechanism which slows and speeds up the angular speed of the cam lobe during its rotation. Arranging the lobe to slow during its open period is equivalent to lengthening its duration. However it has a fundamental drawback in that each individual valve (or each pair of valves in a "four-valve" engine) needs its own eccentric drive and its own controller. This means that to have all the valves in a four-cylinder engine to be able to be varied in duration, eight separate drives and eight separate controllers would be needed. Economics and complexity make this almost impossible. Although the Rover VVC is a somewhat flawed system it is important in that it is the only genuinely variable duration system that has ever been available commercially for a road car – which is a little surprising considering the long history of the internal combustion engine.

Three-Dimensional Cam Lobe. This system consists of a cam lobe that is somewhat elongated axially and has at one end a short duration/reduced lift profile, and at the other end a longer duration/greater lift profile. The two profiles progressively vary in a continuous manner between the two extremes. By shifting the cam lobe axially a stationary follower is exposed to a varying lobe profile to produce different amounts of lift and duration. Although this seems to be a simple and effective idea it has a fundamental (and apparently insoluble) problem which has meant that it has never been used commercially. The varying profile, by its very nature, means that the lobe flanks of the cam lobe are not parallel to the axis of rotation of the camshaft. However the lobe surfaces on the base circle region are parallel to the cam's axis of rotation. This means that a flat-faced follower has to tilt in various directions as the lobe rotates. Even if the follower has freedom to tilt freely it still does not act totally correctly in a geometrical sense on the lobe surface. The only type of follower which is geometrically correct is a spherical or ball-ended type. The resultant point contact stress makes this arrangement generally impractical. Although in popular automotive culture this system is generally attributed to Ferrari, it is concept that dates from the 1930s, or possibly even earlier. It is still being suggested in recent times in various patents (USPTO 7341032 for example). Ferrari has never used this system commercially in any of their road cars.

Combined Cam Lobe Profile Types

This class is divided into three basic subclasses. Two Shaft Combined Profile. This arrangement consists of two closely spaced parallel camshafts of conventional layout. A pivoting follower spans both camshaft and is acted on by two lobes simultaneously. Each camshaft has a phasing mechanism which allows its angular position relative to the engine's crankshaft to be adjusted. The basic idea behind this system is that one lobe controls the opening of a valve and the other controls the closing of the same valve – the spacing of the two events being a period of variable duration. The system is quite workable in that valve clearance, smooth actuation of the valve and other practical issues are not a problem.

However there is something of a basic flaw in that at long duration settings one lobe may be starting to reduce its lift as the other is still increasing – this has the effect of lessening the overall lift and possibly causing dynamic problems. There are also more obvious problems in the overall size of layout due to the parallel shafts, the necessarily large follower etc. Surprisingly, although workable in many aspects, it is not known to have ever been used commercially. The Mechadyne company (from the U.K) has been developing a version of this basic idea having the two shafts arranged coaxially, the lobes attached to the inner and outer shafts being able to be phased separately, a "summation linkage" replacing the conventional follower. This goes someway towards solving the space and follower size problems. Mechadyne also claim to have solved the uneven rate of opening of the valve problem to some extent thus allowing long duration at full lift.

Coaxial Two Shaft Combined Profile. It should be noted here that despite the very similar description to the Mechadyne arrangement the mechanism described here is of a basically different principle. This system is distinctly different to the Mechadyne system in that, although similarly coaxial with pairs of lobes acting on the one follower etc.

the actual lobe profile has a distinctly different profile and acts on a conventional (although maybe somewhat wider than normal) follower. The profile used has a characteristic "snub-nosed" appearance with usually about a twenty degree angular region of true radius about the axis of rotation of the camshaft on the lobe nose. This type of profile is actually very little different from some conventional lobe profiles that act on valve lift multiplying followers of a fairly high ratio of lobe lift to valve lift.

The essential operating principle is that the one follower spans the pair of closely spaced lobes. Up to the angular limit of the nose radius the follower "sees" the combined surface of the two lobes as a continuous, smooth surface. When the lobes are exactly aligned the duration is at a minimum (and equal to that of each lobe alone) and when at the extreme extent of their misalignment the duration is at a maximum. The basic limitation of the scheme is that only a duration variation equal to that of the lobe nose true radius (in camshaft degrees or double this value in crankshaft degrees) is possible. In practice this type of variable cam has a maximum range of duration variation of about forty crankshaft degrees. Any attempt to drastically increase this range (by increasing the angular extent of the true radius lobe nose region) soon gives rise to problems with excessive rates of valve acceleration etc.

This range limits the usefulness of the cam to some extent. Prototypes of the cam system have been run and have proved to be quite reliable and to provide useful peak power increases of the order of thirty to forty percent. Once again it is a little surprising that the general idea has not been used commercially. This system is also notable in being quite often the idea that inventors come up with, at least initially. It is also the principle behind what seems to be the very first variable cam suggestion appearing in the USPTO patent files in 1925 (1527456). The "Clemson Camshaft" referred to elsewhere in this article is of this type. There are literally hundreds of patents in the USPTO files on this general theme alone. A more modern and workable design of this basic principle (with a notably practical and rugged internal structure and neat self-contained centrifugal controller/ actuator) is USPTO 7007652.

Combined Two Shaft Coaxial Combined Profile with Helical Movement. This system is usually referred to more simply as the "Helical Camshaft". It has a very similar principle to the previous type described using virtually (or sometimes exactly) the same base duration lobe profile. However instead of a strictly two-dimensional rotation (that is the rotation is in a single plane) the adjustment is both axial and rotational giving a helical or three-dimensional aspect to its movement. This unusual (and difficult to describe in words) movement overcomes the restricted duration range in the previous type. The duration range is theoretically unlimited but typically would be of the order of one hundred crankshaft degrees.

A range like this would be sufficient for all high-performance needs and more novel matters like engine load control by LIVC (late inlet valve closing) or HCCI (homogenous charge compression ignition). The cam is reportedly somewhat tricky and expensive to make requiring very accurate helical machining and careful assembly.

The helical camshaft does appear to be, of the all the basic classes and subclasses, the one system that could be regarded as, for instance, a car enthusiast's idea of what an ideal variable duration camshaft should be like – a cam that literally has a usable range of variable duration operating through followers of conventional design. It is slightly surprising that it has not been used commercially to date – at least in some sort of specialised application like racing or engine research.

Electro-Mechanical

This method of valve control is often found in camless engine designs. A proponent of this technology is Valeo, which has indicated that its design will be utilised in volume production by 2009.

In this design the valves are opened and closed and held open or closed by means of electromagnets.

Some of the problems which may be encountered with this methodology are:

1. Deceleration of the valve once set in motion is difficult to accomplish. This difficulty is exacerbated by the greatly increased magnetic pull of the magnets as the valves approach the end of their travel (magnetic pull increases dramatically as the distance from the magnet decreases). Inadequate slowing down of the valve can cause significant deterioration of the valve seat and other parts. Utilising springs to effect valve deceleration limits the engine to lower speeds and on its own will not effect a gentle landing of the valve on its seat at all engine speeds.
2. Springs utilised in this type of system may require very careful balancing with the valve movement in order to achieve gentle

valve seating at differing engine speeds. As the springs deteriorate or the engine speed changes, the valve and spring balance may be compromised and ultimately lead to failure.

3. The electromagnets will draw a significant amount of electrical energy, which may require a higher capacity alternator, which will in turn reduce the potential fuel efficiency of the engine.
4. A powerful computer coupled with complex fast-acting control circuitry and devices will likely be necessary to control the valves in real time.

Hydraulic

This type of valve control has been advocated in the search for a camless engine. Sturman Industries, which incorporated its design into a large truck engine a number of years ago, is a proponent of this technology. (The truck did the hill climb at Pikes Peak) Various methods have been explored to utilise hydraulic mechanisms to move the engine valves. Some claim to be successful at low engine speeds, but few claim to achieve that goal meaningfully at the higher RPM requirements of passenger vehicles.

Hydraulic systems suffer from two inherent problems :

1. The faster a liquid is moved, the more it tends to act like a solid. A fast-acting hydraulic system to activate automotive valves at the speeds required in passenger vehicles could require immense pressures, with all the incumbent problems, including the additional energy requirements of the hydraulic pump. Even if higher engine speeds were achieved, valve movement would likely be abbreviated and not fully follow the desired or optimum lift schedule.
2. Temperatures can vary seasonally over a wide range. The hydraulic medium could change viscosity as the temperatures change, which could cause variances in the system's performance which may be difficult to control.

Utilising springs to assist the hydraulic system may also prevent the engine attaining higher speeds. In order to achieve gentle valve seating, hydraulic systems must be carefully controlled. This control may require the use of powerful computers and very precise sensors.

Stepper Motors

This methodology of valve control has previously not been successful in camless engine design due to the limited RPM range inherent in the design.

The following link indicates that some measure of success has been achieved and a camless design for a medium to slow revving engine may be feasible.

Valves that open and close in fixed times cannot optimize engines running at differing speeds and importantly severely restrict engine speed. This is because the degrees of rotation for the valve events increase as the engine speed increases to the point where they are no longer practical. Powertrain claim in their documentation at the above link that their device operates the valves "to 8,000 rev/min and beyond". This figure was likely obtained under laboratory conditions. Real world engines using their devices with 7 ms valve open and close times may have useful speeds limited to well under 6,000 RPM, perhaps even under 5,000 RPM (7 ms at 8,000 RPM is 336 degrees of rotation which is currently unworkable in passenger cars). The engine may be able to run at higher RPM by limiting the lift and hence shortening the valve open close time but will most likely have a lower power output than is achieved at a lower RPM due to the lessening of breathing capability.

Pneumatic

Systems utilising pneumatics to drive the engine valves would in all probability not be feasible because of their complexity and the very large amount of energy required to compress the air.

Cargine Engineering AB, a Swedish Company, has produced pneumatic valves and has fitted them into several different engines. One of these test engines is running in a test Saab 9-5.

The first prototype of the Scuderi Engine uses Cargine pneumatic valves for the intake and exhaust valves.

Implementations

- Aftermarket modifications– Conventional hydraulic tappet can be engineered to rapidly bleed-down for variable reduction of valve opening and duration.
- Alfa Romeo
  - Twin Cam– Latest versions are equipped with Variable Valve Timing technology.
  - Twin Spark – is equipped with Variable Valve Timing technology.
  - JTS — is equipped with Variable Valve Timing technology, both intake and exhaust.

- o Multiair continuously varies the timing of the inlet valve by changing oil pressure.
- BMW
  - o Valvetronic — Provides continuously variable lift for the intake valves; used in conjunction with Double VANOS.
  - o VANOS — Varies intake timing by rotating the camshaft in relation to the gear.
  - o Double VANOS — Continuously varies the timing of the intake and exhaust valves.
- Daihatsu
  - o DVVT — Daihatsu Variable Valve Timing. Continuously varies the timing of the intake camshaft, or both the intake and exhaust camshafts (depending on application).
- Ford
  - o VCT Variable Cam Timing — Varies valve timing by rotating the camshaft.
  - o Ti-VCT Twin Independent Variable Camshaft with two fully variable camshafts used in Ford Sigma engine and Ford Duratec engine.
- Chrysler — Varies valve timing through the use of concentric camshafts developed by Mechadyne enabling dual-independent inlet/exhaust valve adjustment on the 2008 Dodge Viper.
- General Motors Corporation (GM) (Opel/Vauxhall, Chevrolet, GMC, Buick, Cadillac, Holden)
  - o VVT — Varies valve timing continuously throughout the RPM range for both intake and exhaust for improved performance in both overhead valve and overhead cam engine applications.
  - o DCVCP (Double Continuous Variable Cam Phasing) — Varies intake and exhaust camshaft timing continuously with hydraulic vane type phaser; available on Family 1, Family 0, and Family II engines.
  - o Alloytec — Continuously variable camshaft phasing for inlet cams; continuously variable camshaft phasing for inlet cams and exhaust cams (High Output Alloytec).
- Honda
  - o VTEC — Varies duration, timing and lift by switching between two different sets of cam lobes.

- o VTEC-E — This system is designed solely for the purpose of improving fuel economy. A variation of the VTEC mechanism is used to create an offset of lift between the two intake valves, one valve opening only slightly to prevent accumulation of fuel in the intake port. The asymmetrical opening of the intake valves creates a powerful swirl in the combustion chamber and allows for a very lean intake charge to be used under certain conditions. Under normal operation the two intake valve rocker arms are locked together and both valves follow the normal lift cam profile.
- o i-VTEC — In high-output DOHC 4-cylinder engines, the i-VTEC system adds continuous intake cam phasing (timing) to traditional VTEC. In economy-oriented SOHC and DOHC 4-cylinder engines the i-VTEC system increases engine efficiency by delaying the closure of the intake valves under certain conditions and by using an electronically controlled throttle valve to reduce pumping loss. In SOHC V6 engines the i-VTEC system is used to provide Variable Cylinder Management which deactivates one bank of three cylinders during low demand operation.
- o Advanced VTEC — This is the latest Honda VVT system and is the most unusual of all the VTEC systems. Rather than switching between cam lobes the Advanced VTEC system uses intermediate rocker arms with a variable fulcrum to continuously vary intake valve timing, duration and lift.

- Hyundai MPI CVVT — Varies power, torque, exhaust system, and engine response.
- Iran Khodro
  - o CVVT-i - Continuous Variable Valve Timing Intelligence which is used for EF7 & EF4 engines of the IKCO EF engines family.
- Kawasaki — Varies position of cam by changing oil pressure thereby advancing and retarding the valve timing, 2008 Concours 14 (also known as the 1400GTR).
- Lexus VVT-iE — Continuously varies the intake camshaft timing using an electric actuator.
- Mazda S-VT — Continually varies intake timing and crank angle using an oil control valve actuated by the ECU to control oil pressure.

- Mitsubishi MIVEC — Varies valve timing, duration and lift by switching between two different sets of cam lobes.
  - The 4B1 engine series uses a different variant of MIVEC which varies timing (phase) of both intake and exhaust camshafts continuously.
  - The 4N1 engine family is the world's first to feature a variable valve timing system applied to passenger car diesel engines.
- Nissan
  - N-VCT — Varies the rotation of the cam(s) only, does not alter lift or duration of the valves.
  - VVL — Varies timing, duration, and lift of the intake and exhaust valves by using two different sets of cam lobes.
  - CVTCS - introduced with the HR15DE, HR16DE, MR18DE and MR20DE engines in September 2004 on the Nissan Tiida and North American version named Nissan Versa (in 2007); and finally the Nissan Sentra (in 2007). Also used on the new MR16DDT.
  - VVEL - introduced with the VQ37VHR Nissan VQ engine in 2007 on the Infiniti G37.
- Porsche
  - VarioCam — Varies intake timing by adjusting tension of a cam chain.
  - VarioCam Plus — Varies intake valve timing by rotating the cam in relation to the cam sprocket as well as duration, timing and lift of the intake and exhaust valves by switching between two different sets of cam lobes.
- Proton
  - Campro CPS — Varies intake valve timing and lift by switching between two sets of cam lobes without using rocker arms as in most variable valve timing systems. Debuted in the 2008 Proton Gen-2 CPS and the 2008 Proton Waja CPS.
  - Campro CFE — Continuously varies the intake valve timing and cam phasing only, not to be confused with the Campro CPS. Debuted in the 2012 Proton Exora Bold.
  - VVT introduced in the Waja 1.8's F4P renault engine (Toyota supplies the VVT to renault)

- PSA Peugeot Citroën CVVT — Continuous variable valve timing.
- Renault Clio Renault Sport 172, 172 Cup, 182, 182 Cup, Trophy, 197, 197 Cup, 200, and Clio V6 Mk2 VVT — Megane 1.6 vvt variable valve timing. Clio Mk4 Dynamique S 1.6 VVT. RS Twingo 133 1.6 VVT
- Rover VVC — Varies timing with an eccentric disc.
- Suzuki — VVT — Suzuki M engine
- Subaru
  - AVCS — Varies timing (phase) with hydraulic pressure, used on turbocharged and six-cylinder Subaru engines.
  - AVLS — Varies duration, timing and lift by switching between two different sets of cam lobes (similar to Honda VTEC). Used by non-turbocharged Subaru engines.
- Toyota
  - VVT — Toyota 4A-GE 20-Valve engine introduced VVT in the 1992 Corolla GT-versions.
  - VVT-i — Continuously varies the timing of the intake camshaft, or both the intake and exhaust camshafts (depending on application).
  - VVTL-i — Continuously varies the timing of the intake valves. Varies duration, timing and lift of the intake and exhaust valves by switching between two different sets of cam lobes.
  - Valvematic
- Volkswagen Group — VVT introduced with later revisions of the 1.8t engine, and the 30-valve 2.8 L V6. Similar to VarioCam, the intake timing intentionally runs advanced and a retard point is calculated by the ECU. A hydraulic tensioner retards the intake timing. Most modern VW Group petrol engines now include VVT on either the inlet cam, or both inlet and exhaust cams, as in their V6, V8 and V10 engines.
- Volvo
  - CVVT — Continuous variable valve timing on intake and/or exhaust camshafts (depending on application).
  - CPS — Changes valve timing, duration and lift of the intake valves by switching between two different sets of cam lobes. Same basic technology as Porsches VarioCam

Plus with switching direct-acting tappets. To date this is only used on Volvos short inline-6 (SI6) naturally aspirated 3.2 L engine.

- Yamaha — VCT (Variable Cam Timing) Varies position of cam thereby advancing and retarding the valve timing.

Camless

Most four-stroke piston engines today employ one or more camshafts to operate poppet valves. The lobes on the camshafts operate cam followers which in turn open the poppet valves. A camless (or, free valve engine) uses electromagnetic, hydraulic, or pneumatic actuators to open the poppet valves instead. Actuators can be used to both open and close the valves, or an actuator opens the valve while a spring closes it.

As a camshaft normally has only one lobe per valve, the valve duration and lift is fixed. The camshaft runs at half the engine speed. Although many modern engines use camshaft phasing, adjusting the lift and valve duration in a working engine is more difficult. Some manufacturers use systems with more than one cam lobe, but this is still a compromise as only a few profiles can be in operation at once. This is not the case with the camless engine, where lift and valve timing can be adjusted freely from valve to valve and from cycle to cycle. It also allows multiple lift events per cycle and, indeed, no events per cycle—switching off the cylinder entirely.

Camless engines are not without their problems though. Common problems include high power consumption, accuracy at high speed, temperature sensitivity, weight and packaging issues, high noise, high cost, and unsafe operation in case of electrical problems.

Camless valve trains have long been investigated by several companies, including Renault, BMW, Fiat, Valeo, General Motors, Ricardo, Lotus Engineering, Ford and Cargine. Some systems are commercially available, although not in production car engines.

Notably, Formula One cars do not use camless valve trains, but pneumatic valve springs together with conventional camshafts and followers instead, this is however primarily due to the regulations teams must follow for engine development.

Chapter 4

Continuous Variable Valve Timing

Continuous variable valve timing offers a unique ability to have independent control of the intake and exhaust valves in an internal combustion engine. For any engine load criteria, the timing of intake and exhaust can be independently programmed. The main variations of valve timing are: Late intake valve closing, Early Intake valve closing, Early Intake valve opening and early exhaust valve closing. These variations exist on a spectrum, which means engine performance could be optimized under all conditions. Variable Valve Timing has existed since Alfa Romeo implemented it in 1980. The technology took a big step forward with Honda's VTEC system in 1987, which consists of two unique cam profiles for each intake/exhaust valve. The engine can then "decide" based on driving conditions which one to use to maximise performance. This certainly allows a greater degree of control over the engine with obvious performance benefits. However, if valve timing could be controlled independent of crankshaft rotation, then a near infinite number of valve timing scenarios could be accommodated which would dramatically improve fuel economy and emission levels of an automobile.

Traditional Internal Combustion Engines

Traditional Internal Combustion Engines are designed to optimize efficiency and performance, while minimising emissions, under a wide variety of driving conditions. The way this is traditionally accomplished is relatively simple: the driver controls the volume of air allowed into the engine by way of a throttle plate linked to the gas pedal. Electronic sensors in the intake stream measure the intake air volume and temperature, and the Powertrain Control Module (PCM) determines the required fuel amount, as well as the appropriate spark timing to allow the air/fuel mixture to burn optimally and maximise output. Generally, fuel volume and spark timing are pulled off a fuel map – a computer program which looks at various inputs from engine sensors.

The intake and exhaust valves are fixed, meaning they are driven by the engine crankshaft, and they open and close at set intervals during the rotational cycle, independent of engine load conditions. When an engine is designed, the inherent inflexibility in the timing of the valve events results in a compromise.

Simply, engineers must design the engine to run optimally under a wide variety of driving conditions, but perfectly under none. During most driving, the engine is operating at part load and fuel economy is the priority. At part load, the air-fuel ratio is kept stoichiometric to enable efficient operation of the catalytic converter".

One of the major inefficiencies of the Internal Combustion Engine is the loss of rotational energy caused by the intake stroke. As the piston travels downward during the intake stroke, air is drawn into the cylinder via engine vacuum. The pressure is below atmospheric, which means the engine has to do work to take in the air – at the expense of volumetric efficiency.

"These pumping losses depend on the opening and closing of the throttle valve. The losses are high when the throttle valve tends to close and are low at wide open throttle. Thus the pumping losses are inversely proportional with engine load". If intake air volume were to be controlled by the valvetrain, rather than a throttle plate, the pumping losses could be reduced.

"The inlet-valve timing is the most important parameter for optimizing the engine volumetric efficiency, whereas the exhaust-valve timing controls the RGF which reduces exhaust NO emissions. For the engine to operate efficiently and effectively over its entire operating range and conditions, the valve event should be able to vary with speed and load anywhere on the engine map".

Late Intake Valve Closing

The first variation of continuous variable valve timing involves holding the intake valve open slightly longer than a traditional engine. This results in the piston actually pushing air out of the cylinder and back into the intake manifold during the compression stroke. The air which is expelled fills the manifold with higher pressure, and on subsequent intake strokes the air which is taken in is at a higher pressure. Late intake valve closing has shown to reduce pumping losses by 40% during partial load conditions, as well as decreasing NO emissions by 24%. Peak engine torque showed only a 1% decline, and Hydrocarbon emissions were unchanged.

Early Intake Valve Closing

Another way to decrease the pumping losses associated with low engine speed, high vacuum conditions is by closing the intake valve earlier than normal. This involves closing the intake valve midway through the intake stroke. Air/fuel demands are so low at low-load conditions and the work required to fill the cylinder is relatively high, so Early intake valve closing greatly reduces pumping losses. Studies have shown early intake valve closing to reduce pumping losses by 40%, as well as to increase fuel economy by 7%. It also has reduced NO emissions by 24% at partial load conditions. A possible downside to early intake valve closing is that it significantly lowers the temperature of the combustion chamber, which can increase hydrocarbon emissions.

Early Intake Valve Opening

Early intake valve opening is another variation that has significant potential to reduce emissions. In a traditional engine, a process called valve overlap is used to aid in controlling the cylinder temperature. By opening the intake valve early, some of the inert/combusted exhaust gas will back flow out of the cylinder, via the intake valve, where it cools momentarily in the intake manifold. This inert gas then fills the cylinder in the subsequent intake stroke, which aids in controlling the temperature of the cylinder and NO emissions. It also improves volumetric efficiency, because there is less exhaust gas to be expelled on the exhaust stroke.

Early/Late Exhaust Valve Closing

Early and late exhaust valve closing can also reduce emissions. Traditionally, the exhaust valve opens, and exhaust gas is pushed out of the cylinder and into the exhaust manifold by the piston as it travels upward. By manipulating the timing of the exhaust valve, engineers can control how much exhaust gas is left in the cylinder. By holding the exhaust valve open slightly longer, the cylinder is more emptied and ready to be filled with a bigger air/fuel charge on the intake stroke. By closing the valve slightly early, more exhaust gas remains in the cylinder which increases fuel efficiency. This allows for more efficient operation under all conditions.

Drawbacks

The main factor preventing this technology from wide use in production automobiles is the ability to produce a cost effective means of controlling the valve timing under the conditions internal to an engine. An engine cruising at 70 miles per hour (110 km/h) will be doing approximately 3000 revolutions per minute. Every second, the crankshaft rotates 50 times, so the valve timing events have to occur at precise times to offer performance benefits. Problems which need to be overcome before this technology becomes widespread include "dynamic performance in the high engine speed range, power consumption, long term durability and repeatability". Electromagnetic valve actuators offer the greatest control of precise valve timing, but are not cost effective for production vehicles at this time. Currently, automakers use a variety of different mechanical techniques to continuously vary valve timing.

Cam Phasing Variable Valve Timing

Cam-phasing VVT is the simplest, cheapest and most commonly used mechanism at this time. This system varies the valve timing by shifting the phase angle of the camshafts. For example, at high engine speeds, the intake camshaft will be rotated to advance the intake valve timing. This movement is controlled by the engine management system according to need, and actuated by hydraulic cam gears. It is important to note that cam-phasing variable valve timing cannot vary the duration of valve opening. It simply allows earlier or later valve opening. Earlier opening results in earlier closing, of course. It also cannot vary the valve lift. Cam-phasing variable valve timing is the simplest and cheapest method because each camshaft needs only one hydraulic actuator, unlike other systems that employ individual mechanism for every cylinder. "The greatest degree of variability in the valvetrain is achieved when each individual valve has its own timing system, enabling it to be opened and closed as and when required". Obviously, actuating each valve individually offers the most precise control over the valve timing, but the cost is often prohibitive.

Applications

Automakers have been testing this technology for years, as its benefits are obvious. Several automakers have begun implementing this technology in recent years, but its widespread use is still to come. Several automakers, including BMW, Nissan, and Toyota currently offer models with some form of continuous variable valve timing. Debuted in the BMW 316ti Compact in 2001, Valvetronic was the first continuous variable valve lift mechanism made into production. The goal of Valvetronic was to reduce fuel consumption, rather than to maximise power. Valvetronic regulates engine output by varying the depth of valve lift according to the position of throttle pedal. This means the conventional throttle butterfly can be disabled which reduces pumping loss. BMW achieved 10% reduction in fuel consumption with Valvetronic. BMW's Valvetronic adds an electric motor, an eccentric shaft and an intermediate rocker arm at each intake valve. The intake camshaft acts on the intermediate rocker arms through roller bearings. When the driver calls for more power, the electric motor turns the eccentric shaft, which pushes the intermediate rocker arms and in turn pushes the valve to open deeper.

Nissan introduced its Variable Valve Event and Lift (VVEL) in 2007 as the world's second CVVL system. The first application was on the VQ37VHR V6 engine of Skyline Coupe (Infiniti G37). Nissan's system is more compact than BMW's, involve less parts and energy loss, and is therefore more adaptable to high-performance engines. The VVEL does not use a conventional intake camshaft. Each valve is actuated by a cam which is pivoted on - but not fixed to - the camshaft. While conventional camshafts feature fixed rotating lobes, the cam in VVEL swings up and down; this is why it does not need a symmetric profile. Its movement is driven by the camshaft via a series of component. VVEL varies valve lift by the eccentric control shaft inside the rocker arm. By rotating the eccentric control shaft, the position of rocker arm is shifted, changing the swing angle of cam. The swing angle of cam determines the degree of valve lift.

Toyota introduced its Valvematic technology in 2008. Valvematic employs an intermediate shaft to achieve continuous variable valve lift. The intermediate shaft has an actuating member for each cylinder. Each actuating member is made of two finger followers on either side of a roller bearing member. The followers can rotate in relation to the roller member by means of internal gear threads and an electric motor attached to the end of the intermediate shaft. The gear threads of the roller member and finger followers are in opposing directions. This means when the shaft swivels, the roller member and finger followers will move in opposing directions, moving either apart or closer together. In this way, the axle angle between them can be varied infinitely by the electric motor. The intake valve is actuated by camshaft via the intermediate shaft. Specifically, the camshaft acts on the roller member of intermediate shaft, transferring the movement to both finger followers, then towards the roller rocker arms and ultimately to the intake valves. When the angle of the finger followers is narrow with respect to the roller member, valve lift is low. As the angle increases, so does valve lift. In this way, Valvematic can vary valve lift by adjusting the angle of the finger followers in relation to the roller bearing member.

Zone Valve

A zone valve is a specific type of valve used to control the flow of water or steam in a hydronic heating or cooling system.

In the interest of improving efficiency and occupant comfort, such systems are commonly divided up into multiple zones. For example, in a house, the main floor may be served by one heating zone while the upstairs bedrooms are served by another. In this way, the heat can be directed principally to the main floor during the day and principally to the bedrooms at night, allowing the unoccupied areas to cool down.

This zoning can be accomplished in one of two ways:

- Multiple circulator pumps, or
- A single circulator pump and zone valves.

Zone valve construction and operation

Zone valves as used in home hydronic systems are usually electrically powered. In large commercial installations, vacuum or compressed air may be used instead. In either case, the motor is usually connected to the water valve via a mechanical coupling.

For electrical zone valves, the motor is often a small shaded-pole synchronous motor combined with a rotary switch that can disconnect the motor at either of the two stopping points ("valve open" or "valve closed"). In this way, applying power to the "open valve" terminal causes the motor to run until the valve is open while applying power at the "close valve" terminal causes the motor to run until the valve is closed. The motor is commonly powered from the same 24 volt ac power source that is used for the rest of the control system. This allows the zone valves to be directly controlled by low-voltage thermostats and wired with low-voltage wiring. This style of valves requires the use of an SPDT thermostat or relay.

Zone valves can also be constructed using wax motors and a spring-return mechanism. In this case, the valve is normally closed by the force of the spring but can be opened by the force of the wax motor. Removal of electrical power re-closes the valve. This style of zone valve operates with a perfectly ordinary SPST thermostat.

For vacuum- or pneumatically-operated zone valves, the thermostat usually switches the pressure or vacuum on or off, causing a spring-loaded rubber diaphragm to move and actuate the valve. Unlike the electrical zone valves, these valves automatically return to the default position without the application of any power, and the default position is usually "open", allowing heat to flow.

Highly sophisticated systems may use some form of building automation such as BACnet or LonWorks to control the zone valves.

Comparison with Multiple Circulator Pumps

Multiple zones can be implemented using either multiple, individually-controlled circulator pumps or a single pump and multiple zone valves. Each approach has advantages and disadvantages.

Multiple Pump System

Advantages:

- Simple mechanical and control design ("SPST thermostats")
- Redundancy: If one zone pump fails, the others can remain working

Disadvantages:

- Higher initial installation cost. Circulator pumps cost more than zone valves
- Circulating pumps have a shorter life span than a zone valve
- Higher power consumption. Operating circulators draw substantial power any time the zone is actively heating. Zone valves, by comparison, draw little power at any time and many designs only draw power while in transition from open to close or vice-versa.

Zone Valve System

Advantages:

- Lower initial installation cost.
- Lower power consumption.
- Ease of maintenance certain modals.

Disadvantages:

- Zone valves can be unreliable. Zone valves operated by electric timing motors aren't "fail safe" (failing to the "open" condition).
- No inherent redundancy for the pump. A zone-valved system is dependent upon a single circulator pump. If it fails, the system becomes completely inoperable.
- The system can be harder to design, requiring both "SPDT"

thermostats or relays and the ability of the system to withstand the fault condition whereby all zone valves are closed simultaneously.

Air-operated Valve

An Air Operated Valve is a type of power operated valve that uses air pressure against a piston or diaphragm to produce linear or circular movement to operate a valve. Air Valves perform two important functions in a piping system. They maintain system design efficiency and provide system protection. System efficiency is maintained by venting air via the air valves from the system that can restrict flow and increase pumping costs. Protection is provided by exhausting and admitting air through the air valves during system operations including start-up, shutdown, and critical conditions such as power failures or line breaks. The exhausting and admitting of air during these conditions will reduce the potential for destructive surges and water hammer normally associated with uncontrolled air or a vacuum condition within the piping system.

Angle Seat Piston Valve

An angle seat piston valve is a pneumatically-controlled valve with a piston actuator providing linear actuation to lift a seal off its seat. The seat is set at an angle to provide the maximum possible flow when unseated. Angle seat piston valves are particularly suited to applications where high temperatures and large flowrates are required, such as steam or water. When used in reverse some models of angle seat piston valve will eliminate water hammer when operated.

Operation

Pneumatic valves are operated by a pilot medium under pressure, usually compressed air but also oil or water.

The valve is equipped with a pneumatic actuator supplied by a three-way solenoid valve. The pressure of the pilot medium enters the actuator cylinder and acts on the piston, which allows the seal to open or to close through the stem. The return of the seal into its rest position is usually achieved by a return spring that can be found in the pneumatic actuator.

In the double-acting configuration there is no return spring and the pilot medium is used both for opening and for closing the valve.

A red indicator becomes visible through the sight dome that is found on the top of the actuator when the valve is in the open position.

The pressure range depends on the pressure of the pilot medium and the controlled medium and on the direction of the flow, but also on construction parameters of the valve, such as the diametre of the orifice, the diametre of the actuator cylinder and the spring force.

Beer Tap

A beer tap is a valve, specifically a tap, for controlling the release of beer. While in other contexts, depending on location, a "tap" may be a "faucet", "valve" or "spigot", the use of "tap" for beer is almost universal. This may be because the word was originally coined for the wooden valve in traditional barrels. Beer served from a tap is largely known as draught beer, though beer served from a cask is more commonly called cask ale, while beer from a keg may specifically be called keg beer.

Usage

Pressure-dispense Bar Tap

Almost universally in modern times, bulk beer is supplied in kegs that are served with the aid of external pressure. In a normal bar dispense system, this pressure comes from a cylinder of carbon dioxide (or occasionally nitrogen) which forces the beer out of the keg and up a narrow tube to the bar. At the end of this tube is a valve built into a fixture (usually somewhat decorative) on the bar. This is the beer tap, and opening it with a small lever causes beer, pushed by the gas from the cylinder, to flow into the glass.

Portable Keg Tap

Sometimes, beer kegs designed to be connected to the above system are instead used on their own, perhaps at a party or outdoor event. In this case, a self-contained portable tap is required that allows beer to be served straight from the keg. Because the keg system uses pressure to force the beer up and out of the keg, these taps must have a means of supplying it. The typical "picnic tap" uses a hand pump to push air into the keg; this will cause the beer to spoil faster but is perfectly acceptable if the entire keg will be consumed in a short time. Portable taps with small CO_2 cylinders are also available.

Cask Beer Tap

Beers brewed and served by traditional methods, typically cask ale, do not use artificial gas. Taps for cask beer are simple on-off valves that are hammered into the end of the cask. When beer is served directly from the cask ("by gravity"), as at beer festivals and some pubs, it simply flows out of the tap and into the glass. When the cask is stored in the cellar and served from the bar, as in most pubs, the beer line is screwed onto the tap and the beer is pulled through it by a beer engine.

The taps used are the same, and in beer-line setups the first pint is often poured from the cask as for "gravity", for tasting, before the line is connected. Cask beer taps can be brass (now discouraged for fear of lead contamination), stainless steel, plastic, and wood.

In Scotland, cask ale was traditionally served through a *tall fount*. These appear similar to keg taps (indeed, many Scottish pubs serve keg beer through adapted tall founts) rendered from brass but the beer was drawn from the barrel via air pressure generated by a water engine rather than by a pressurised artificial gas. Some pubs still dispense cask beer using this method (although the water engine will invariably have been replaced by an electric compressor) but is increasingly rare due to the perception that hand-pumps are the 'correct' means by which to serve real ale, and to the potential for confusion with keg ale, which is regarded as inferior by many beer drinkers.

Dimensions

Typical tap handles are 9-13" tall, but can range from very short (4") to novelty lengths (>13"). Most taps are assembled with a 4" delrin handle.

The threading for tap handles is 3/8"-16 UNC (coarse). The faucet is male and the handle is female.

Bjork–Shiley Valve

The Bjork–Shiley valve is a mechanical heart valve prosthesis. Beginning in 1971, it has been used to replace the aortic or mitral valves. It marks the first example of a successfully used tilting-disc valve. It was manufactured first by Shiley Inc., then later by Pfizer after that company purchased Shiley. One model of the Bjork–Shiley valve became the subject of a famous lawsuit and recall after it was shown to malfunction.

Construction

The Bjork valve consists of a single carbon-coated disc in a tantalum housing. The discs are held in place by two metal struts, an inflow and an outflow strut. The standard design Bjork–Shiley valve is a very durable valve and was widely used in the 1970s. According to the literature the housing is actually an alloy called Haynes 25 which is a chromium cobalt alloy.

Convexo-Concave Valves

Attempts to improve the design of the valve and speed manufacturing, however, led to a weaker structure, with serious consequences. Beginning in 1979, Bjork–Shiley valves with the convexo-concave design had a tendency to develop fractures in the outflow strut which could result in catastrophic valve failure and possibly sudden cardiac death. Later analysis revealed that the strut was fracturing at the place where it was welded onto the metal valve ring. One end of the strut would fracture first, followed by the second strut some months later. Eventually, 619 of the 80,000 convexo-concave valves implanted fractured in this way, with the patient dying in two thirds of those cases. The convexo-concave valve was forced off the market by the FDA in 1986. Valves welded by specific welders were at greater risk of fracture. The Convexo-Concave valve was withdrawn from the market and multiple lawsuits filed. This recall and legal struggle have been described as "perhaps the most infamous recall case on record".

Not all of these valves were removed and replaced. The risk of valve fracture must be balanced against the risk of the surgery to replace the valve. Decision tools have been published and are available from a trust set up for this purpose during the litigation surrounding the valve "recall". The tools take into account patient age, the size and location of the valve (aortic or mitral) and other factors to come up with a recommendation as to whether surgery to replace the valve is worthwhile. In some patients the risk of surgery to replace the valve is higher than the risk of the valve fracturing. A compensation fund was also established to provide money for patient compensation and to pay for the costs of heart surgery to replace the valve.

More recent models of the Bjork–Shiley valve did away with the convexo-concave design and are reliable valves that do not require replacement.

Brass Instrument Valve

Brass instrument valves are valves used to change the length of tubing of a brass instrument allowing the player to reach the notes of various harmonic series. Each valve pressed diverts the air stream through additional tubing, individually or in conjunction with other valves. This lengthens the vibrating air column thus lowering the fundamental tone and associated harmonic series produced by the instrument. Valves in brass instruments require regular maintenance and lubrication to ensure fast and reliable movement.

Piston Valve

The piston valve found in the majority of valved brass instruments was invented in 1838 and patented in 1839. They are sometimes called Périnet valves after their inventor, François Périnet. They work by diverting air obliquely through ports in the stock of the valve so that a loop of tubing is included in the air stream, lowering the pitch. The stock of the valve is cylindrical and moves up and down through a larger cylindrical casing.

Rotary Valve

The rotary valve works using a short circular plug housed in a larger cylindrical casing. Ports are cut along one plane through the stock and direct the airflow into the lengths of tubing when the stock is rotated 90 degrees. Trombone F-attachment valves are usually rotary, with several variations on the basic design also in use, such as the Thayer axial flow valve and Hagmann valve. French horns almost always have rotary valves, and many tubas have rotary valves. Joseph Riedlin is credited with the first use of rotary valves on brass instruments in 1832.

Axial Flow Valve

The Thayer axial flow valve is a replacement for the traditional rotary valve found on trombones with F attachments. Invented by Orla Ed Thayer (with assistance from Zig Kanstul) in 1976. It uses a conical plug, allowing a deflection of the airflow by only 25 degrees or less.

Double-Piston Valve

The double-piston valve, also called the Vienna valve or pumpenvalve, is a type of valve that preceded the modern single piston Périnet valve. It was first produced in a trumpet in 1821 by Christian Friedrich Sattler of Leipzig. In this valve type, the simultaneous movement of two pistons bends the air flow in two right angles to introduce an additional valve loop. These turns cause constrictions in the bore, that make the instrument harder to play, but they are still used today despite their drawbacks. At first, the two pistons were operated by a lever connected with braces, but the later Vienna model of these valves was operated by long rods connecting the pistons to spring-loaded keys on the other side of the instrument.

They are called "Vienna valves" because while they have fallen out of favour in comparison with the modern rotary valve in almost all places, they are still used almost exclusively in Vienna, Austria, where players prefer the more smooth legato and mellow, natural horn-like tone.

Stölzel Valve

The Stölzel valve bears the name of its inventor, Heinrich Stölzel, who first applied these valves to the french horn in 1814. Until that point, there had been no successful valve design, and horn players had to stop off the bell of the instrument, greatly compromising tone quality to achieve partial music. In a Stölzel valve, the air enters through the bottom of the valve casing, up through the hollow bottom end of the piston, and through a port to the valve loop. The air is then led through an oblique port in the piston to a short tube connecting the valves where it is then directed through the second valve and out the bottom. This type of valve, however, had inherent problems. It forced the air to double back on itself and the 90 degree turns disrupted the bore, causing significant undesired back-pressure. These problems were improved upon later by the double-piston valve.

Clutch Delay Valve

A clutch delay valve is an automotive component added to the clutch system of an automobile to standardize clutch engagement speed. Its purpose is to engage the drive train of an automobile without introducing shock to the drive train components by engaging too quickly. By preventing drive train shock, CDVs also prevent an automobile's balance from being upset, which aids in handling characteristics.

Clutch delay valves operate on the principle that engagement only needs to occur at a certain rate to be efficient and that engaging at a quicker rate can damage the drive train components of an automobile. Engaging the drive train to the engine too quickly can damage drive train parts, including the transmission, differential, half shafts, axles, and CV joints. Engaging the drive train too slowly can damage the clutch friction disc and cause clutch slippage.

Clutch delay valves operate as one-way restrictor valves that limit the volume of fluid that can move through the lines in a given time. This ensures a steady quick and firm engagement without being too fast for the drive train to handle since clutches can quickly disengage but re-engagement is damped.

Cusps of Heart Valves

The cusps of the heart valves serve to seal the heart valves when closed. There are three cusps for each valve except for the mitral valve, which has only two (hence its alternate name, "bicuspid valve").

"Nodules" are located at the tip of the valve, to form a tighter seal.

Semilunar valves

- Pulmonary valve
  - o "Right cusp"
  - o "Left cusp"
  - o "Anterior cusp".

Nowhere in the literature does it describe the echocardiographic location of each pulmonary valve. According to researcher and professor Dr. Paul Michael, in the RVOT view the anterior and right cusps can be viewed.

- aortic valve. (Note that one is absent in the condition bicuspid aortic valve.)
  - o "Right cusp"
  - o "Left cusp"
  - o "Posterior cusp" (or "noncoronary cusp").

Atrioventricular valves

- Tricuspid valve- right av
  - o "Anterior cusp" (or "infundibular cusp")
  - o "Posterior cusp" (or "marginal cusp")
  - o "Septal cusp" (or "medial cusp")
- Mitral valve - left av
  - o "Anterior cusp"
  - o "Posterior cusp".

Double Dump Valve

Double dump valves, also known as double flap valves or double flap gates, are a type of airlock valve commonly used in industrial applications as a component in bulk material handling applications. Double dump valves are primarily used to discharge chunky or fibrous, bulk materials from hoppers, bins, and cyclones operating under positive or negative pressure. Double dump valves are used to discharge a flow of material while at the same time serving as an airlock transition point to preserve the pressure differential above and below the valve. This type of material handling valve is ideal for use with bulky or abrasive materials that would tend to jam or damage a rotary feeder.

Operation

Double dump valves have two trap-door flap plates which open and close on an alternating cycle. The top gate opens to allow collected material to drop through onto the bottom gate. The top gate then re-closes to prevent air leakage above the valve. The bottom gate then opens to allow material to pass completely through the valve. The bottom gate then re-closes to prevent air leakage below the valve. This cycle is continuously repeated in order to maintain a steady discharge of material from the valve, while maintaining a positive seal on the system.

Design

Timing of the flap plates is critical in order to make sure a double dump valve operates properly. There are several designs that have been employed in order to maintain a consistently operating valve.

Control systems

Pneumatic: Each flap plate has its own pneumatically operated cylinder and solenoid valve. The timing of the valve is controlled by an electronic signal sent to the solenoid valves. This method of operation is the most expensive and allows for accurate timing of the valve.

Counterbalance

This type of double dump valve works essentially, like a stacked pair of trickle valves. Each flap plate is held closed with a counterweighted arm. As material builds up on the top plate the weight of the head of material pushes it open where it dumps onto the bottom plate. This type of double dump valve has no electrical or pneumatic controls and is the least expensive method of operation. However, it is difficult to sustain a properly timed cycle.

Cam-operated

A motor-driven cam rotates and alternately opens the top gate and the bottom plate. A spring attached to the gate's shaft re-closes the gates. This type of double dump valve is generally the most reliable and can be operated with an electric motor or a pneumatic motor. The speed of the cycle can be selected through the gear reducer ratio or by adjusting the motor speed.

Dustcap

A dustcap is a small cover used on the valve stem of a bicycle or car tire to prevent dust or other small particles from entering the valve and damaging it. The dustcap also forms a pressure seal, helping to prevent deflation of the tire due to slight gas seepage past the tire valve. Dustcaps are usually made from plastic but may occasionally be made of metal. The dustcap is internally threaded and is secured by screwing it onto the end of the stem. To prevent corrosion due to electrolysis and malfunction of the electronics in a direct TPMS and avoid costly repairs, metal valve caps without insulating plastic linings should not be used on direct-TMPS-equipped tire valve stems.

A dustcap may also refer, on a bicycle, to the crankarm bolt cover. It is usually a piece of metal or plastic that, on the crankarm, is snapped or screwed into the threaded counterbore, which houses the bolt that holds the crankarm to its axle, the bottom bracket spindle. The dustcap serves to protect the threads (on the inside of the counterbore), which are used with a crankarm puller to remove the crankarm from the spindle. It is also possible for the crank arm to squirm on the bottom bracket spindle, causing the crankarm nut/bolt to become loose, leading to the nut/bolt working its way out of the bottom bracket spindle. The dust cap limits the amount of dangerous loosening that can occur.

A dustcap may also refer, on a bicycle, to the metal ring that covers the bearings in a cup-and-cone ball-bearing hubset. These types of dust caps, being press-fit, are difficult to remove with bending or damaging them. It is suggested to press them out from the other side of the hub using a screw driver. In this role the dustcap reduces the dust that enters the bearing races, and extends service intervals.

Chapter 5

Electrohydraulic Servo Valve

An Electro-Hydraulic Servo Valve or EHSV is an electrically operated valve that controls how hydraulic fluid is ported to an actuator. Servo valves and Servo-Proportional Valves are electro-hydraulic operate by transforming a changing analogue or digital input signal into a smooth set of movement in a hydraulic cylinder. Servo valves can provide precise control of position, velocity, pressure and force with good post movement damping characteristics.

Control

A low voltage is used control the servo valve. The control voltage is passed into an amplifier which provides the power to alter the valve's position. The valve will then deliver a measured amount of fluid power to an actuator. The use of a feedback transducer on the actuator returns an electrical signal to the amplifier to condition the strength of the voltage to the servo valve.

Moog Servo Valves

Many of the servo valves in use today are manufactured by Moog Inc. A typical example of a servo valve is the Moog servo valve D631 series. An example of servo valve use is in Blow Molding where the servo valve controls the wall thickness of extruded plastic making up the bottle or container by use of a deformable die.

Fuse (Hydraulic)

In hydraulic systems, a fuse is a component which prevents the sudden loss of hydraulic fluid pressure. It is a safety feature, designed to allow systems to continue operating, or at least to not fail catastrophically, in the event of a system breach. It does this by stopping the flow of hydraulic fluid through itself if the flow exceeds a threshold. The term "fuse" is used here in analogy with electrical fuses which perform a similar function.

Hydraulic systems rely on high pressures (usually over 7000 kPa) to work properly. If a hydraulic system loses fluid pressure, it will become inoperative and components such as actuators may collapse. This is an undesirable condition in life-critical systems such as aircraft or heavy machinery. Hydraulic fuses help guard against catastrophic failure of a hydraulic system (for instance, by line breakage or component failure) by automatically isolating the defective branch.

When a hydraulic system is damaged, there is generally a rapid flow of hydraulic fluid towards the breach. Most hydraulic fuses detect this flow and seal themselves if the flow exceeds a predetermined limit. There are many different fuse designs but most involve a passive spring-controlled mechanism which closes when the pressure differential across the fuse becomes excessive.

Types

There are two types of hydraulic fuses. The first one acts like a pressure relief valve, venting in case of a pressure surge. The second is more or less like a check valve. The only difference is a check valve is in place to prevent upstream fluid from coming back and venting out. A fuse is in place before the venting area and stops fluid from venting forward of it. Hydraulic fuses are not a perfect solution to fluid loss. They will probably be ineffective against slow, seeping loss of hydraulic fluid, and they may be unable to prevent fluid loss in the event of a catastrophic system failure involving multiple breaches to hydraulic lines. Also, when a fuse activates it is likely that the system will no longer function as designed. Depending on the system, hydraulic fuses may reset automatically after a delay, or may require manual re-opening.

In Dam Spillways

In the design of a spillway for a dam, a fuse plug is a water retaining structure designed to wash out in a controlled fashion if the main dam is in danger of overtopping due to flood, and if the normal spillway channel is insufficient to control the over topping.

Guardian Valve

A guardian valve is a valve used in marine steam turbine propulsion plants to prevent steam from leaking into the astern turbine while the vessel is operating in the ahead mode. It is normally installed between the astern throttle valve and the astern elements of the low pressure (LP) turbine. Typically only the LP turbine of a steam ship's propulsion plant has reversing blade elements. Steam leaking in such a manner would result in a loss of efficiency and possibly overheat and damage the turbine blades. The guardian valve must be opened prior to any maneuvering situation, in order to permit the astern turbine to be used to bring the vessel to a rapid stop or to back down. Any similar arrangement of a stop valve used to protect leaking past a throttling valve could also be referred to as a guardian valve.

Hagmann Valve

The Hagmann Valve is a rotary valve for trombone that was developed by René Hagmann in the early 1990s. His intention was to address some of the problems with the popular Thayer Valve, in particular the maintenance issues caused by its relatively complex design. A standard rotary valve turns the air flow as much as 90 degrees when it is engaged. This does not substantially affect the sound of instruments like the French Horn which already have lots of turns in the air flow, but part of the characteristic sound of the trombone comes from its long, straight air flow. The angles on a Hagmann valve are dramatically less (60 to 66 degrees) which makes it much easier for the player to maintain proper tone with the valve engaged. While the Hagmann Valve does not achieve as low an angle as the Thayer Valve (which can be as low as 20 degrees), it is much easier to maintain and the tone differences are nearly inaudible.

Many leading trombone manufacturers now offer Hagmann valves as an option, especially on their professional ranges.

Hydrogen Valve

A hydrogen valve is a special type of valve that is used for hydrogen at very low temperatures or high pressures in hydrogen storage or for example hydrogen vehicles.

Types

High pressure ball valves up to 6000 psig (413 bar) at 250 degrees F (121 degrees C) and flow coefficients from 4.0 to 13.8.

Material

Valves used in industrial hydrogen and oxygen applications, such as petrochemical processes, are often made of inconel.

Isolation Valve

An isolation valve is a valve in a fluid handling system that stops the flow of process media to a given location, usually for maintenance or safety purposes. They can also be used to provide flow logic (selecting one flow path versus another), and to connect external equipment to a system. A valve is classified as an isolation vavle because of its intended function in a system, not because of the architecture of the valve itself. Therefore, many different types of valves can be classified as isolation valves.

To easily understand the concept of an isolation valve, one can think of the valves *under* a kitchen or bathroom sink in a typical household. These valves are normally left open so that the user can control the flow of water with the spigot above the sink, and does not need to reach under the counter to start or stop the water flow. However, if the spigot needs to be replaced (i.e. maintenance needs to take place on the system), the isolation valves are shut to stop the flow of water when the spigot is removed. In this system, the isolation valves and the spigot may even be the same type of valve. However, due to their function they are classified as the isolation valves and, in the case of the spigot, the control valves.

Common Applications

- Firewater control
- Pipeline safety systems
- Residential plumbing systems (both water and gas)
- Nuclear reactors
- Oil and gas wells.

Oral Inflation Valve

An oral inflation valve is present on most swim toys and beach balls. Since the valve is placed in the mouth and saliva can be ejected into the inflatable object, the valve is subjected to some health concerns.

Partial Stroke Testing

Partial stroke testing (or PST) is a technique used in a safety instrumented system to allow the user to test a percentage of the possible failure modes of a shut down valve without the need to physically close the valve.

Pilot-operated Relief Valve

Like other pressure relief valves (PRV), pilot operated relief valves (PORV) are used for emergency relief during overpressure events (e.g. a tank gets too hot and the expanding fluid increases the pressure to dangerous levels). The distinction between PORV and conventional PRV is that pilot valves use system pressure to seal the valve. A PRV typically uses a spring to hold the disc or piston on seat. The essential parts of a PORV are a pilot valve (or control pilot), a main valve, a pitot tube, the dome, a disc or piston, and a seat. The volume above the piston is called the dome. PORV are also called POSV (Pilot Operated Safety Valve), POPRV (Pilot Operated Pressure Relief Valve), or POSRV (Pilot Operated Safety Relief Valve), depending on the manufacturer and the application. Technically POPRV is the most generic term, but PORV is often used generically (as in this article) even though it should refer to valves in liquid service.

Mode of Functioning

The pressure is supplied from the upstream side (the system being protected) to the dome often by a small pitot tube. The downstream side is the pipe or open air where the PORV directs its exhaust. The outlet pipe is typically larger than the inlet. 2"x3", 3"x4", 4"x6", 6"x8", 8x10" are some common sizes.

The upstream pressure tries to push the piston open but it is opposed by that same pressure because the pressure is routed around to the dome above the piston. The area of the piston exposed to pressure is larger in the dome than it is on the upstream side; the result is a net sealing force. F = PA (Force = Pressure * Area)

The pressure from the pitot tube to the dome is routed through the actual control pilot valve. There are many designs but the control pilot is essentially a conventional PRV with the special job of controlling pressure to the main valve dome. When the pilot valve reaches set pressure it opens and releases the pressure from the dome. The piston is then free to open and the main valve exhausts the system fluid. The control pilot opens either to the main valve exhaust pipe or to atmosphere.

Snap Acting: At set pressure the valve snaps to full lift, it can be quite violent on large pipes with significant pressure. The pressure has to drop below the set pressure in order for the piston to reseat.

Modulating: The pilot is designed to open gradually, so that less of the system fluid is lost during each relief event. The piston lifts in proportion to the overpressure. Blowdown is typically short.

Comparison to Conventional Pressure Relief Valves

Advantages:

- Smaller package on the larger pipe sizes.
- More options for control.

- Seals more tightly as the system pressure approaches but does not reach set pressure.
- Control pilot can be mounted remotely.
- Some designs allow for changes in orifice size within the main valve.

Disadvantages:

- More complex.
- More expensive at smaller sizes (starts to even out as pipe size increases).
- Small parts in pilot valve are sensitive to contaminant particles.

Plastic Pipework

Plastic pipework is used for the conveyance of drinking water, waste water, chemicals, heating and cooling fluids, foodstuffs, ultra-pure liquids, slurries, gases, compressed air and vacuum system applications.

Materials Used:

- ABS (Acrylonitrile Butadiene Styrene)
- UPVC (Unplasticized Polyvinyl Chloride)
- CPVC (Post Chlorinated Polyvinyl Chloride)
- PP (Polypropylene)
- PE (Polyethylene) also known as LDPE, MDPE and HDPE (low, medium, and high density)
- PVDF (Polyvinylidene fluoride).

Material Characteristics

ABS (Acrylonitrile Butadiene Styrene): ABS is used for the conveyance of potable water, slurries and chemicals. UPVC has excellent chemical resistance across its operating temperature range, with a broad band of operating pressures. Due to its long-term strength characteristics, high stiffness and cost effectiveness, UPVC systems account for a large proportion of plastic piping installations. CPVC is resistant to many acids, bases, salts, paraffinic hydrocarbons, halogens and alcohols.

PP (Polypropylene)

Polypropylene is suitable for use with foodstuffs, potable and ultra pure waters, as well as within the pharmaceutical and chemical industries.

PE (Polyethylene)

Polyethylene has been successfully used for the safe conveyance of potable and waste water, hazardous waste, and compressed gases for many years.

PVDF (Polyvinylidene Fluoride)

PVDF has excellent chemical resistance which means that it is widely used in the chemical industry as a piping system for aggressive liquids.

Components of Plastic Pressure Pipe Systems

Pipes, fittings, valves, and accessories make up a plastic pressure pipe system. The range of pipe diametres for each pipe system does vary. However, the size ranges from 12 to 400 mm (0.472 to 15.748 in) and D_8 to 16 in (9.53 to 406.40 mm).

Pipes are extruded and are generally available in: 3 m (9.84 ft), 4 m (13.12 ft), 5 m (16.40 ft), and 6 m (19.69 ft) straight lengths and 25 m (82.02 ft), 50 m (164.04 ft), 100 m (328.08 ft), and 200 m (656.17 ft) coils for LDPE and MDP.

Pipe fittings are moulded and come in many sizes: Tee 90° equal (straight and reducing), Tee 45°, Cross equal, Elbow 90° (straight and reducing), Elbow 45°, Short Radius Bend 90° Socket/coupler (straight and reducing), Union, End Caps, Reducing Bush, and Stub, Full face, and blanking flanges.

Valves are moulded and also come in many types: Ball Valve (also multiport), Butterfly valve, spring-, ball-, and swing-check non-return valves, Diaphragm valve, Knife Gate Valve, Globe Valve and Pressure Relief/Reduction Valves

Pneumatic Non-return Valves

Pneumatic non-return valves are used where a normal non-return valve would be ineffective. This is for example where there is a risk of flood water entering a site but an equal risk of pollution or a chemical spills leaving a site and polluting the environment.

Pneumatic non-return valves are installed below ground and can be used to pneumatically lock the non-return valve closed thus containing a site in the event of a spill. It is common practice to lock sites using pneumatic non-return valves during the loading or transferring of chemicals or hazardous waste. Pneumatic non-return valves have a longer service life when compared to pneumatic bladder systems.

Proportioning Valve

A proportioning valve is a valve that relies on the laws of fluid pressure to distribute input forces to one or more output lines. A proportioning valve can increase or decrease forces for each output, depending on the cross-sectional surface areas of those output lines.

A simple example is an input tube with cross-sectional area A entering a chamber. Leading out of the chamber are two more tubes, one with cross-sectional area $3A$ and the other with area $A/2$. If a force F is applied to the fluid in the input tube, the pressure in that tube will be F/A. Utilising pressure laws, we find that each output tube will see the same pressure. This means the output tube with area $3A$ will yield a force of $3F$, and the output tube with area $A/2$ will yield a force of $F/2$. Thus, if you apply a 10-pound force to the input, you will get forces of 30 lbf and 5 lbf, respectively, from the outputs. If you apply a 10-newton force to the input, you will get forces of 30 N and 5 N, respectively, from the outputs

Proportioning valves are frequently used in cars to reduce the brake fluid pressure to the rear brakes.

Relay Valve

A relay valve is an air-operated valve. It is used in air brake systems to remotely control the brakes at the rear of a heavy truck or semi-trailer in a tractor-trailer combination. Relay valves are necessary in heavy trucks in order to speed-up rear-brake application and release, since air takes longer to travel to the rear of the vehicle than the front of the vehicle, where the front service brakes, foot-valve, parking-control valve, and trailer-supply valve (if applicable) are located.

Without relay valves, it would take too long for sufficient air to travel from the brake pedal valve to the rear of the truck or trailer in order to apply the rear service brakes concurrently with the front service brakes, resulting in a condition known as brake lag. To correct this condition on a long-wheel-base vehicle, a relay valve is installed near the rear service brake chambers. In tractors as well as straight-trucks, a remote air-supply is provided in the form of a large-diametre pipe connected between the primary reservoir and the relay valve for remote service brake application.

Trailers

In trailers, this remote air-supply is in the form of a tank, which is charged whenever the emergency brakes are released via the red trailer-supply valve on the dashboard. In a dual-circuit air brake system, this tank actually receives its air from both the primary and secondary reservoirs of the tractor; the air from both of these reservoirs is merged via a two-way check valve.

The two-way check valve is a pneumatic device that has two inputs and one output; each input is connected to one these reservoirs. Only the air that is at the higher pressure is allowed to pass through to the check valve's output, which then passes through the tractor-protection valve, and then travels onward towards the trailer's air-tank and spring brake valve via the red trailer-supply line (a.k.a., the emergency line); this releases the trailer's emergency brakes (a.k.a. spring brakes). The tractor-protection valve is a device that prevents air from being lost from the tractor's braking system in the event of the air-lines becoming separated or broken. The tractor's air-lines connect to the trailer's air-lines via metal connectors known as gladhands. The merged air from both reservoirs of the tractor prevents air-loss from only one tractor braking circuit from causing the trailer's spring brakes to automatically apply. This gives the driver more control, and prevents the vehicle from grinding to a halt in an unsafe location, such as in the middle of an intersection.

Service Brake Relay Valve

With a service brake relay valve installed, the hose that connects to the primary delivery-port output of the foot-valve becomes a control-line (i.e., The air from the foot-valve "dead ends" at the relay valve's control-port.). Only low-volume air-signals are required to travel back and forth between the foot-valve's delivery port and the relay valve's control port; therefore, the air-volume supplied by the delivery port is now only a tiny-fraction of what otherwise would have been required had the relay valve not been installed.

This reduces the delay between the application of the front and rear brakes to only a fraction of a second. When the driver depresses the brake pedal, a small amount of air momentarily opens the relay valve's supply port, which then directs air from the remote air-supply directly to the rear service brake chambers, and quickly applies the rear service brakes. The pressure delivered to the service brake chambers in this manner will equal the control-pressure delivered by the foot-valve to the relay valve. When the driver partially or fully releases the brake pedal, the control-pressure delivered by the foot-valve decreases; this causes the relay valve's supply port to close, and its exhaust port to momentarily open, thus preventing a pneumatic short-circuit from occurring while the air exhausts from all rear service brake chambers. The air exhausts until the pressure within the service brake chambers once again equals the control-port pressure, which partially or fully releases the rear service brakes.

In order to control the trailer service brakes, the merged outputs (i.e., merged via 2 two-way check valves connected in-series to give three inputs) of the foot-valve and trailer-hand-valve (if applicable) are directed through the tractor-protection valve, and onward towards the trailer relay valve via the blue service line. In tractors that are not equipped with a trailer hand valve, only the merged outputs of the foot-valve (i.e., via a single two-way check valve) are directed towards the trailer relay valve; however, the fact that the foot-valve's delivery-port outputs are still merged enables the trailer's service brakes to still be controlled even if there is failure within one braking circuit of the tractor.

Spring Brake Relay Valve

A spring brake relay valve works on the same principle as the service brake relay valve, although it has the opposite effect. This type of relay valve responds to a major drop in pressure at its control-port by opening its exhaust port, which causes the air from each spring brake chamber under its control to remotely exhaust, thus applying the spring brakes much-more-quickly than would otherwise be possible if the air were required to discharge via the yellow parking-control valve on the dashboard. In a dual-circuit air brake system, air from both the primary and secondary reservoirs is fed into the supply-port of the parking-control valve, as well as the supply-port of this relay valve; it is merged via yet another two-way check valve. The delivery-port output of the parking-control valve connects to the control-port of this relay valve; this enables the spring brakes to be controlled via this valve.

The merged air from the parking-control valve prevents air-loss from only one braking circuit from causing the spring brakes to automatically apply. This gives the driver more control, and prevents the vehicle from grinding to a halt in an unsafe location. However, with this increased control, comes increased responsibility on the part of the driver: If air is lost from the primary circuit alone, the spring brakes must be manually applied by the driver via the parking-control valve; otherwise, the front service brakes may not be enough to stop the vehicle safely in an emergency—especially if the vehicle is heavily loaded, and/or travelling at a high-speed. In fact, the driver's failure to manually apply the spring brakes in this situation could lead to catastrophic failure of the front brakes due to overheating, since it could cause the front service brakes to exceed their design-limit for energy absorption.

The relay valve's function is analogous to the transistor used in electronic circuits.

Reversing Valve

A reversing valve a type of valve and is a component in a heat pump, that changes the direction of refrigerant flow. By reversing the flow of refrigerant, the heat pump refrigeration cycle is changed from cooling to heating or vice versa. This allows a residence or facility to be heated and cooled by a single piece of equipment, by the same means, and with the same hardware.

Operation

The reversing valve has two states, relaxed and energized. The energized state is typically achieved by applying 24 volts of alternating current, which is commonly used in HVAC. The heat pump can be designed by the manufacturer to produce cooling or heating with the reversing valve in the relaxed state. When the reversing valve is energized, it will produce the opposite conditioning. In other words, a reversing valve installed in such a way as to produce cooling when relaxed will produce heating when energized. Likewise, a reversing valve installed to produce heating when relaxed will produce cooling when energized.

Control

Depending on the construction and use of the heat pump, the reversing valve may be driven by the heat pump through the use of a control board, or it may be driven directly by a thermostat (typically from the "O" terminal).

Replacement

Reversing valves are built into the heat pump by the manufacturer, and must be replaced by an HVAC technician if they fail.

Rotolock Valve

A Rotolock valve is a special refrigeration valve with a teflon ring seated against a machined surface enclosed by a threaded fitting. It is used in many applications of the refrigeration industry, mainly when specifically requested by the customer. This valve is easily installed and removed in only seconds. When in place, this valve will allow the technician to remove all refrigerant from the compressor without requiring removal of the system charge. This provides a huge time savings as well as a huge monetary savings. Not only is money saved by reducing labour costs, but the need for replacement refrigerant is minimised. If the compressor is part of a large system the system charge could amount to several thousand dollars.

Safety Shutoff Valve

Safety shutoff valves are safety valves used to close a line and stop the flow of material. Some types are manually closed; others are automated to act when a preset condition (such as a failure in the system) occurs.

Principle

When this type of valve is tripped, the flow is quickly stopped and an indicator disc tells the operator that the electrical circuit has been opened by a failure somewhere in the system. When the system failure has been corrected so that the circuit is again closed, the valve can be opened by the action of the hand lever. If, however, the system failure has not been satisfactorily corrected, the circuit will remain open and moving the hand lever will not open the valve, because the valve stem remains disengaged from the handle.

In a solenoid-operated safety shutoff valve, a spring action closes the valve instantly when an electric current fails and the solenoid ceases to be energized. The solenoid circuit is generally arranged so that it is broken upon failure of any element of the system. This valve cannot be re-opened until the solenoid is again energized.

A safety shutoff valve should be fail-safe, that is close upon failure of any element of the input control system (such as temperature controllers, steam pressure controllers), air pressure, fuel pressure, current from a flame detector, or current from other safety devices such as low water cutoff, and high pressure cutoff.

However, sudden closing of a valve in a piping system may lead to water hammer or implosion so in special cases there may be additional items connected to the shutoff valve, such as a pressure relief valve or an aerator valve. The coil of the valve solenoid must be connected in series with all of the elements. For this to operate, fuel, air and steam pressure can be converted to electrical signals by means of bellows, a bourdon tube, or a diaphragm-operated mercury switch.

Shut Down Valve

A shut down valve (also referred to as SDV or Emergency shutdown valve, ESV, ESD, or ESDV) is an actuated valve designed to stop the flow of a hazardous fluid or external hydrocarbons (gases) upon the detection of a dangerous event. This provides protection against possible harm to people, equipment or the environment. Shutdown valves form part of a Safety instrumented system. The process of providing automated safety protection upon the detection of a hazardous event is called Functional Safety. Shutdown valves are primarily associated with the petroleum industry although other industries may also require this type of protection system. ESD valves are required by law on any equipment placed on an offshore drilling rig to prevent catastrophic events like the BP Horizon explosion in the Gulf of Mexico in 2010.

Types of Valve

For fluids, metal seated ball valves are used as shut-down valves (SDV's). Use of metal seated ball valves leads to overall lower costs when taking into account lost production and inventory, and valve repair costs resulting from the use of soft seated ball valves which have a lower initial cost. Straight-through flow valves, such as rotary-shaft ball valves, are typically high-recovery valves. High recovery valves are valves that lose little energy due to little flow turbulence. Flow paths are straight through. Rotary control valves, butterfly valve and ball valves are good examples. For air intake shut down, two distinct types are commonly utilised, i.e. butterfly valves and swing gate or guillotine valves. Because diesel engines ignite fuel using compression instead of an electronic ignition, shutting off the fuel source to a diesel engine will not necessarily stop the engine from running. When an external hydrocarbon, such as methane gas, is present in the atmosphere, it can be sucked into a diesel engine causing overspeed or over revving, potentially leading to a catastrophic failure and explosion. When actuated, ESD valves stop the flow of air and prevent these failures.

Types of Actuation

As shutdown valves are form part of a SIS it is necessary to operate the valve by means of an actuator. These actuators are normally fail safe fluid power type. Typical examples of these are:

- Pneumatic cylinder
- Hydraulic cylinder
- Electro-hydraulic actuator.

In addition to the fluid type actuators also vary in the manner in which the energy is stored to operate the valve on demand as follows:

- Single-acting cylinder - Or spring return where the energy is stored by means of a compressed spring
- Double-acting cylinder - Energy is stored using a volume of compressed fluid.

The type of actuation required depends upon the application, site facilities and also the physical space available although the majority of actuators used for shutdown valves are of the spring return type due to the fail safe nature of spring return systems

Measuring Performance

For shutdown valves used in safety instrumented systems it is essential to know that the valve is capable of providing the required level of safety performance and that the valve will operate on demand. The required level of performance is dicated by the Safety integrity level (SIL). In order to adhere to this level of performance it is necessary to test the valve. There are 2 types of testing methods available being

- Proof test - A manual test that allows the operator to determine whether the valve is in the "as good as new" condition by testing for all possible failure modes and requires a plant shutdown
- Diagnostic Teset - An automated on-line test that will detect a percentage of the possible failrue modes of the shudown valve. An example of this for a shutdown valve would be a partial stroke test

Sleeve Valve (Water)

A sleeve valve is a device typically used in a water supply or distribution system when there is a need to reduce high pressure or throttle flow. The sleeve valve consists of a cylindrical gate (tube) which slides over an inner sleeve. The inner sleeve was consists of a series of nozzles specifically sized and arranged to provide a solution to cavitation issues frequently encountered in throttling applications with other valve types. Because of its design, it minimises cavitation. Cavitation that does occur in severe throttling conditions is directed from the valve down to centre of the valve or pipe so that no erosion damage to the piping can take place.

Types

There are three main types of sleeve valves: inline, angle pattern and downflow pattern. Inline sleeve valve can be installed in the pipeline or at the end of the pipeline for free discharge. Inline sleeve valve have an inlet section which allows upstream media to flow into an outer annular area. The flow then takes a 90° turn through the sleeve and collides with flow from the opposite side dissipating the energy. The flow then travels downstream through an exit spool on the downstream end. The angle pattern sleeve valve is installed at a 90° bend in place of an elbow. Angle pattern sleeve valves employ a similar design as the inline sleeve valve by creating an outer annular area to contain the media and a sleeve with tapered nozzles to pass the flow. The difference is that the flow direction after passing through the sleeve changes by 90°. Downflow pattern sleeve valves are installed at the end of a pipeline and discharge into sumps, reservoirs, stilling wells or tanks. Unlike inline and angle pattern the discharge of a downflow pattern is outward and is not contained by the valve itself.

Advantages

Sleeve valves can operate for long periods of time over a high differential pressure across the valve. When a cavitation condition does exist the cavitation takes place surrounded by water. Unlike other valves, air is not necessary to control or suppress the cavitation. Because of this a single sleeve valve can replace several standard valves used in series. Sleeve valves can control flow and/or pressure over the entire stroke of the valve. Conventional control valves can typically control flow or pressure in the 20% to 80% range. Additionally, unlike conventional control valves the sleeve valve increases flow incrementally, i.e. 10% increase in stroke leads to a 10% increase in flow. Sleeve valves are adaptive to system or consumer requirements. The sleeve valve is designed to meet the requirements and allow for a wide range of variability of function. Most conventional control valves are designed to a specific set of conditions.

Disadvantages

Sleeve valve are generally more expensive than other types of control valves. Typically, sleeve valves are employed in situations where they would replace multiple valves in series or where frequent maintenance may be required, Sleeve valves need to be used where large debris is not expected in the system. As the orifice holes are relatively small, it is possible and likely that any trash in system will collect in front of sleeve or even in the nozzles themselves. Access ports are generally provided to access the valve interior for debris removal.

Installations

Typical installations for sleeve valves include the following: reservoir discharge, pump control, pressure regulation, turbine bypass, and tank level control.

- Reservoir Discharge – Sleeve valves are used to control flow and dissipate excess energy from a reservoir outlet. Commonly used in areas with a high pressure drop, the sleeve valve can discharge to the atmosphere, or to a submerged outlet in a downstream creek.
- Pressure Regulation – A sleeve valve can be used to reduce pressure from a high pressure supply to a lower pressure distribution zone, or to discharge into a tank or reservoir as greatly reduced pressure to minimise damage to the tank. Typically a SCADA system controls the valve position.
- Turbine Bypass – When a valve is needed to bypass an energy recovery turbine, a sleeve valve can minimise upstream and downstream pressure surges by slowly opening and closing in proportion with the turbine. The sleeve valve will dissipate the energy normally consumed by the turbine, thus allowing maintenance to be performed on the turbine, and also maintain downstream flow requirements.nk Level Control – Sleeve valves are ideal to control the water level in a distribution system storage tank in potable water applications.

Slide Check

A slide check is a valve used in pressurized systems that lets air out ahead of it (the destination) while keeping pressure intact from behind it (the source). Slide checks are commonly used in paintball, mostly in remote propulsion setups, to cut air off from the gun while keeping the remote pressurized. Slide checks are also useful when working with high pressure systems, such as hydro testing or audible air or CO_2 leaks as you can rapidly apply pressure and release it without the frighting "pow" or projection of hoses with QD male {quick disconnect} fittings causing a dangerous hose whip motion.

Sun Valve

A sun valve (*Swedish: solventil*, "solar valve") is a form of flow control valve, notable because it earned its inventor Gustaf Dalén the

Nobel prize in physics. The valve formed part of the Dalén light which was used in lighthouses from the 1900s through to the 1960s by which time electric lighting came to dominate. Many prominent contemporary engineers, including Thomas Edison, initially doubted that the device could work. The patent office demanded a special demonstration before approving the patent application.

Design

The valve is controlled by four metal rods enclosed in a glass tube. The central rod that is blackened is surrounded by the three polished rods. As morning sunlight light fell onto all of the rods, the absorbed heat of the sun allowed the unequally expanding dark rod to cut the gas supply. After sunset, the central rod cooled down, becoming the same length as polished rods and opening the gas supply. The gas was lit by the small, always-burning pilot light.

Transmission Solenoid

A transmission solenoid is an electro-hydraulic valve that controls fluid flow into and throughout an automatic transmission. Solenoids can be normally open or normally closed. They operate via a voltage or current supplied via the transmission computer or controller. Transmission solenoids are usually installed in a transmission valve body, Transmission Control Unit, or transmission control module.

Types

- Variable Force Solenoid (VFS)
- On-Off Solenoid
- Pulse-Width Modulated Solenoid (PWM)
- Low Leak Variable Bleed Solenoid.

Trickle Valve

Trickle valves, also known as vacuum valves, are commonly used in industrial dust collection applications to maintain an airlock seal on a dust collector hopper while allowing bulk solid material to be automatically discharged. These valves are typically a less expensive alternative to more commonly used rotary airlocks. Unlike rotary airlock valves which are driven by either an electric motor or a gas or air-powered motor, trickle valves require no external power source.

Operation

Trickle valves can be used only on systems operating under negative pressure. The vacuum created inside the dust collector holds the valve closed, allowing a head of material to build up above the trickle valve. Once the pressure of the material equals the negative pressure of the system, the valve is forced open and a trickle of material will begin flowing from the valve at the same rate it is collected. The magnitude of the vacuum of the system and the bulk density of the material being handled determine how much material will build up before the valve begins to discharge. Because trickle valves work automatically and require no power source or controls, they have very low initial cost and virtually no operating cost.

Design

A well designed trickle valve has few or no moving parts. Most trickle valves are also adjustable as to how much vacuum pressure it takes to operate. Adjustments are typically made with a weighted counterbalance or an adjustable leaf spring. Because these valves have no way to force clogs through, they are only recommended for use with dry and relatively dense, free-flowing materials such as sand, fly ash, and cement.

Types

Duckbill Sleeve Type

This type of trickle valve utilises a specially designed duckbill rubber sleeve. The negative pressure of the dust collection system collapses the sleeve on itself, creating an airtight seal. Adjustable leaf springs are inserted into the sides of the sleeve which help to resist the vacuum. The weight of collected material above the valve forces the sleeve open to allow a steady discharge. Duckbill sleeve trickle valves have no moving parts.

Trap Door Type

Trap door trickle valves have one moving part, a hinged flap plate to maintain a seal between the hopper and the discharge. This type of valve is adjusted with an external weighted counterbalance or it is not adjustable. Greased bearings are required for smooth operation of the flap plate's hinge.

Vacuum Delay Valve

A vacuum delay valve is a valve with a small orifice, which delays a vacuum signal. These are commonly used in automobiles to alter the behaviour of vacuum switches, vacuum motors, and other vacuum devices. The engine in a common automobile produces almost 20 inches of mercury (68 kPa) of vacuum, and this pressure differential has been utilised for everything from automatic locks and windshield wipers to operating emissions control items. Engine vacuum is also the best direct source of information on the engine's load.

Most delay valves have a one-way function, where there is either no restriction or no movement in one direction.

Delay valves are usually colour-coded to their function.

Color Code		*Time (seconds)*	*Threshold (+/- seconds)*	*Direction*
Orange	Black	2	0.5	Forward
Purple	Black	4	0.8	Forward
Gray	Black	10	2	Forward
Brown	Black	20	4	Forward
White	Black	63.5	13.5	Forward
Yellow	Black	100	20	Forward
Green	Black	200	40	Forward
Orange	White	2	0.5	Reverse
Purple	White	4	0.8	Reverse
Gray	White	10	2	Reverse
Gold	White	15	3	Reverse
Brown	White	20	4	Reverse
Yellow	White	100	20	Reverse
Red	White	375	75	Reverse

Valve Actuator

Actuators are used for the automation of industrial valves and can be found in all kinds of technical process plants: they are used in waste water treatment plants, power plants and even refineries. This is where they play a major part in automating process control. The valves to be automated vary both in design and dimension. The diametres of the valves range from a few inches to a few metres.

Depending on their type of supply, the actuators may be classified as pneumatic, hydraulic, or electric actuators.

Classification of Actuators According to Their Movement

Travel means the distance the closing element within the valve has to cover to completely open or close that valve. Typical closing elements include butterfly, globe or gate valve discs. These three closing elements stand for the three basic movements required for covering the travel. The butterfly valve disc is operated by a 90° swivel movement from end position OPEN to CLOSED, the globe valve disc is operated by a rather short linear movement (stroke) while the gate valve disc movement covers the full diametre of the valve. Each movement type requires a specific actuator type.

Multi-turn Actuators

Multi-turn actuators are required for the automation of multi-turn valves. One of the major representatives of this type is the gate valve. The basic requirements on multi-turn actuators are described in the standard EN ISO 5210 as follows:

"A multi-turn actuator is an actuator which transmits to the valve a torque for at least one full revolution. It is capable of withstanding thrust."

A valve stem is mounted to the gate valve disc. The multi-turn actuator moves the gate valve disc from OPEN to CLOSED and vice versa via a stem nut. To cover the complete valve travel, the so-called valve stroke, the actuator has to perform – depending on the valve – a few or several hundred rotations. Due to their design, the stroke of electric actuators, contrary to that of their pneumatic counterparts, has no limits. Therefore, gate valves are exclusively automated by means of electric multi-turn actuators.

The multi-turn actuator has to be able to withstand the weight of the gate valve disc by means of the valve attachment, the interface to the valve. This is expressed in the second sentence of the definition.

Gate valves may have a diametre of approx. 4 inches to several metres. The torque requirement for multi-turn solutions ranges from approx. 10 N m to 30,000 N m.

Part-turn Actuators

Part-turn actuators are required for the automation of part-turn valves. Major representatives of this type are butterfly valves and ball valves. The basic requirements on part-turn actuators are described in the standard EN ISO 5211 as follows:

"A part-turn actuator is an actuator which transmits a torque to the valve for less than one full revolution. It need not be capable of withstanding thrust."

Less than one full revolution usually means a swivel movement of 90°; however, there are some valve types requiring a different swing angle, such as two-way valves. The closing elements in part-turn actuators are always supported by the valve housing, i.e. the weight of the closing element does not act upon the part-turn actuator. This is expressed in the second sentence of the definition.

Part-turn valves diametres range from a few inches to several metres. The torque requirement for operating the closing element has a comparable range from approximately 10 N m to several 100,000

N m. Electric actuators are unrivalled for large-diametre valves with high torque requirements.

Linear Actuators

Currently there is no international standard describing linear actuators or linear thrust units. A typical representative of the valves to be automated is the control valve. Just like the plug in the bathtub is pressed into the drain, the plug is pressed into the plug seat by a stroke movement. The pressure of the medium acts upon the plug while the thrust unit has to provide the same amount of thrust to be able to hold and move the plug against this pressure.

Most of the linear actuators used are pneumatic diaphragm actuators. They are characterised by a simple design principle and are therefore cost-effective. A compressed air supply is a prerequisite for their use. In case this is not possible, the use of thrust units is recommended which can easily be supplied with power.

Design

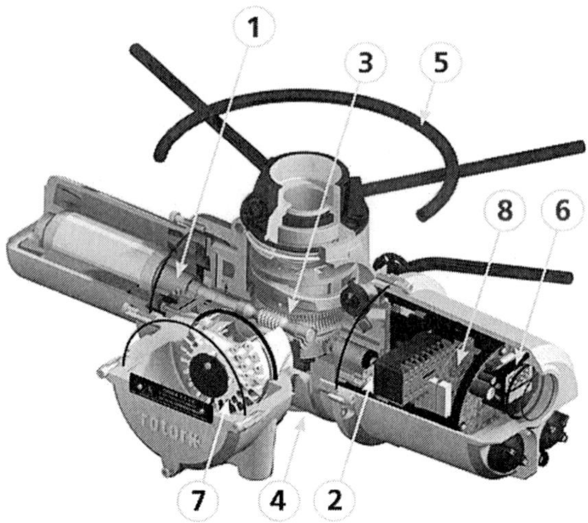

Figure: Electric multi-turn actuator with controls

Motor

Robust asynchronous three-phase AC motors are mostly used as the driving force, for some applications also single-phase AC or DC motors are used. These motors are specially adapted for valve automation as they provide higher torques from standstill than comparable conventional motors, a necessary requirement to unseat sticky valves. The actuators are expected to operate under extreme ambient conditions, however they are generally not used for continuous operation since the motor heat buildup can be excessive.

Limit and Torque Sensors

The limit switching measures the travel and signals when an end position has been reached, the torque switching measures the torque present in the valve. When exceeding a set limit, this is signalled in the same way. Actuators are often equipped with a remote position transmitter which indicates the valve position as continuous current or voltage signal.

Gearing

Often a worm gearing is used to reduce the high output speed of the electric motor. This enables a high reduction ratio within the gear stage, leading to a low efficiency which is desired for the actuators. The gearing is therefore self-locking i.e. it prevents accidental and undesired changes of the valve position by acting upon the valve's closing element. This is of major importance for multi-turn actuators which are axially loaded with the weight of the gate valve disc.

Valve Attachment

The valve attachment consists of two elements.

First: The flange used to firmly connect the actuator to the counterpart on the valve side. The higher the torque to be transmitted, the larger the flange required.

Second: The output drive type used to transmit the torque or the thrust from the actuator to the valve shaft. Just like there is a multitude of valves there is also a multitude of valve attachments.

Dimensions and design of valve mounting flange and valve attachments are stipulated in the standards EN ISO 5210 for multi-turn actuators or EN ISO 5211 for part-turn actuators. The design of valve attachments for linear actuators is generally based on DIN 3358.

Manual Operation

In their basic version most electric actuators are equipped with a handwheel for operating the actuators during commissioning or power failure. The handwheel does not move during motor operation.

Actuator Controls

Both actuator signals and operation commands of the DCS are processed within the actuator controls. This task can in principle be assumed by external controls, e.g. a PLC. Modern actuators include integral controls which process signals locally without any delay. The controls also include the switchgear required to control the electric motor. This can either be reversing contactors or thyristors which, being an electric component, are not subject to mechanic wear. Controls use the switchgear to switch the electric motor on or off depending on the signals or commands present. Another task of the actuator controls is to provide the DCS with feedback signals, e.g. when reaching a valve end position.

Electrical Connection

The supply cables of the motor and the signal cables for transmitting the commands to the actuator and sending feedback signals on the actuator status are connected to the electrical connection. The electrical connection can be designed as a separately sealed terminal bung or plug/socket connector. For maintenance purposes, the wiring should be easily disconnected and reconnected.

Fieldbus Connection

Fieldbus technology is increasingly used for data transmission in process automation applications. Electric actuators can therefore be equipped with all common fieldbus interfaces used in process automation. Special connections are required for the connection of fieldbus data cables..

Functions

Automatic Switching Off in the End Positions

After receiving an operation command, the actuator moves the valve in direction OPEN or CLOSE. When reaching the end position, an automatic switch-off procedure is started. Two fundamentally different switch-off mechanisms can be used. The controls switch off the actuator as soon as the set tripping point has been reached. This is called limit seating. However there are valve types for which the closing element has to be moved in the end position at a defined force or a defined torque to ensure that the valve seals tightly. This is called torque seating. The controls are programmed as to ensure that the actuator is switched off when exceeding the set torque limit.

Safety Functions

The torque switching is not only used for torque seating in the end position, but it also serves as overload protection over the whole travel and protects the valve against excessive torque. If excessive torque acts upon the closing element in an intermediate position, e.g. due to a trapped object, the torque switching will trip when reaching the set tripping torque. In this situation the end position is not signalled by the limit switching. The controls can therefore distinguish between normal operation torque switch tripping in one of the end positions and switching off in an intermediate position due to excessive torque.

Temperature sensors are required to protect the motor against overheating. For some applications by other manufacturers, the increase of the motor current is also monitored. Thermoswitches or PTC thermistors which are embedded in the motor windings mostly reliably fulfil this task. They trip when the temperature limit has been exceeded and the controls switch off the motor.

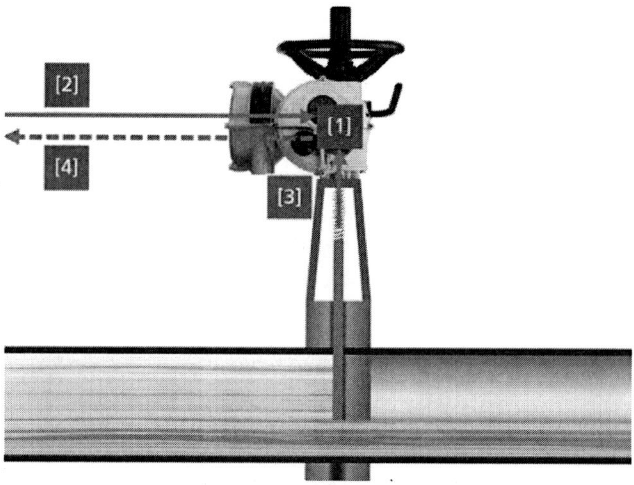

Figure: The positioner is supplied with a setpoint and an actual value [3]. The motor is controlled until the actual value is identical to the setpoint. The DCS generally needs a feedback signal.

Process Control Functions

Due to increasing decentralisation in automation technology and the introduction of micro processors, more and more functions have been transferred from the DCS to the field devices. The data volume to be transmitted was reduced accordingly, in particular by the introduction of fieldbus technology. Electric actuators whose functions have been considerably expanded are also affected by this development. The simplest example is the position control. Modern positioners are equipped with self-adaptation i.e. the positioning behaviour is monitored and continuously optimised via controller parameters.

Meanwhile, electric actuators are equipped with fully-fledged process controllers (PID controllers). Especially for remote installations, e.g. the flow control to an elevated tank, the actuator can assume the tasks of a PLC which otherwise would have to be additionally installed.

Diagnosis

Modern actuators have extensive diagnostic functions which can help identify the cause of a failure. They also log the operating data. Study of the logged data allows the operation to be optimised by changing the parameters and the wear of both actuator and valve to be reduced.

Duty Types

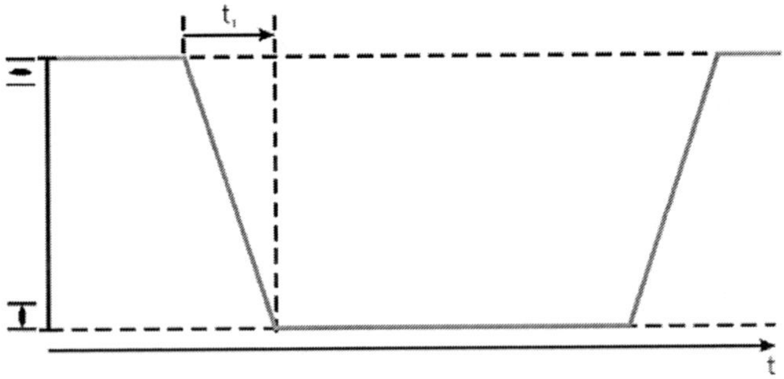

Figure: Typical time period in open-close duty. t1 is the operating time and may not exceed the maximum permissible running time

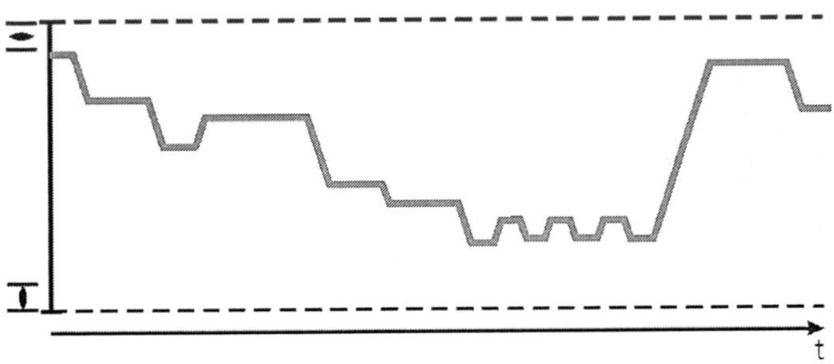

Figure: Typical time period in modulating duty.

Open-close Duty

If a valve is used as a shut-off valve, then it will be either open or closed and intermediate positions are not held. Shut-off valves are rarely operated, the interval between operations may be a few minutes or even several months.

The 'Short-time duty S2' operation mode of the electric motor in accordance with the IEC 34-1 standard indicates that the actuator is suitable for this kind of applications. Another characteristic of this duty type is the maximum permissible running time without interruption. A typical time for actuators is 15 min.

Positioning Duty

Defined intermediate positions are approached for setting a static flow through a pipeline. The same running time limits as in open-close duty apply.

Modulating Duty

The most distinctive feature of a closed-loop application is that changing conditions require frequent adjustment of the MOV e.g. to set a certain flow rate. Sensitive closed-loop applications require adjustments within intervals of a few seconds. The demands on the actuator are higher than in open-close or positioning duty. Both mechanics and motor have to be designed as to be able to withstand the high number of starts without any deterioration in control accuracy.

The duty type of the electric motors suitable for this application is called *intermittent duty S4* or *intermittent duty S5*. The running time is limited by the relative on-time; for modulating actuators this is usually 25 %.

Service Conditions

Electric actuators are used worldwide, in all climate zones, in all kinds of industrial plants under special local ambient conditions. The applications are often safety related, therefore the plant operators put high demands on the reliability of the devices. Failure of an actuator may cause accidents in process-controlled plants and toxic substances may leak into the environment.

Process-controlled plants are often operated for several decades which justifies the higher demands put on the lifetime of the devices.

For this reason, actuators are always designed in high enclosure protection. The manufacturers put a lot of work and knowledge into corrosion protection.

Enclosure Protection

The enclosure protection types are defined according to the IP codes of EN 60529. The basic versions of most electric actuators are designed to the second highest enclosure protection IP 67. This means they are protected against the ingress of dust and water during immersion (30 min at a max. head of water of 1 m). Most actuator manufacturers also supply devices to enclosure protection IP 68 which provides protection against submersion up to a max. head of water of 6 m.

Ambient Temperatures

In Siberia, temperatures down to – 60 °C may occur, and in technical process plants + 100 °C may be exceeded. Using the proper lubricant is crucial for full operation under these conditions. Greases which may be used at room temperature can become too solid at low temperatures for the actuator to overcome the resistance within the device. At high temperatures, these greases can liquify and lose their lubricating power. When sizing the actuator, the ambient temperature and the selection of the correct lubricant are of major importance.

Explosion Protection

Electric actuators are used in applications where potentially explosive atmospheres may occur. This includes among others refineries, pipelines, oil and gas exploration or even mining. When a potentially explosive gas-air-mixture or gas-dust-mixture occurs, the actuator must not act as ignition source. Hot surfaces on the actuator as well as ignition sparks created by the actuator have to be avoided. This can be achieved by a flameproof enclosure, where the housing is designed to prevent ignition sparks from leaving the housing even if there is an explosion inside.

Actuators designed for these applications, being explosion-proof devices, have to be qualified by a test authority (notified body). There is no worldwide standard: depending on the country where the actuators are used, different directives and regulations have to be observed. Within the European Union, ATEX 94/9/EC applies, in US, the NEC (approval by FM) or the CEC in Canada (approval by the CSA). Explosion-proof actuators have to meet the design requirements of these directives and regulations.

Additional Uses

Small electric actuators can be used in a wide variety of assembly, packaging and testing applications. Such actuators can be linear, rotary, or a combination of the two, and can be combined to perform work in three dimensions. Such actuators are often used to replace pneumatic cylinders.

Valve Leakage

Valve leakage refers to flow through a valve which is set in the 'off' state.

The importance of valve leakage depends on what the valve is controlling; a dripping tap is less significant than a leak from a six-inch pipe carrying high-pressure radioactive steam.

ANSI specifies six different leakage classes, with 'leakage' defined in terms of the full open valve capacity

- Class I, or 'dust-tight' valves, are intended to work but have not been tested
- Class II valves have no more than 0.5% leakage with 50psi (or less if operating pressure is less) of air pressure at the operating temperature
- Class III valves have no more than 0.1% leakage under those conditions; this may require soft valve seats, or lapped metal surfaces
- Class IV valves have no more than 0.01% leakage under those conditions; this tends to require multiple graphite piston rings or a single Teflon piston ring, and lapped metal seats.
- Class V valves leak less than 5*10^-12 cubic metres, per second, per bar of pressure differential, per millimetre of port diametre, of water when tested at the service pressure.

Class VI valves are slightly different, in that they have soft seats, and are required (at 50psi or operating pressure whichever is less) to have less than a specified leakage rate in millilitres of air per minute:

1 inch	0.15 ml/min
1.5 inch	0.30 ml/min
2 inch	0.45 ml/min
2.5 inch	0.60 ml/min
3 inch	0.90 ml/min
4 inch	1.70 ml/min
6 inch	4.00 ml/min
8 inch	6.75 ml/min
10 inch	9 ml/min
12 inch	11.5 ml/min

Valve Oil

Valve oil is a lubricant for valves of brass instruments. It is typically mostly mineral oil with a small amount of other ingredients.

Use of Valve Oil

Besides lubricating the moving parts of the valve, valve oil provides corrosion protection to the bare metal of the inner valve. While the valve piston or rotor is made of a metal which is more resistant to corrosion, the inner valve casing is typically bare brass. (The brass on the outside of a modern instrument is lacquered to prevent corrosion.) The oil also completes the seal between the valve casing and the piston or rotor.

Although a clean and unoiled valve of a well maintained instrument should move without unusual force, the inside of a musical instrument is a very inhospitable environment for a delicate valve mechanism. The musician constantly blows warm moist air through the valve. To make matters worse, impurities may be blown from the musician's mouth into the instrument. Even if nothing grows in the valve, the condensation and changing temperature of the metal can cause an untreated valve to bind, possibly resulting in a stuck valve. Even a minor binding of a valve affects the playability of the instrument and is at least very annoying. Also, woodwind musicians use valve oil (called key oil for woodwinds since they do not have valves, they have keys) to lubricate the mechanism of the keys to improve the springback action.

Types of Valve Oil

There are two main types of valve present on brass instruments: piston and rotor. Accordingly, vendors sell different types of oil, including scented varieties. Some vendors recommend up to three different types of oil for some valve types. Slide oil for trombones is also a very similar solution. The difference between the types of oil is primarily the viscosity. The minerals in the valve oil are dangerous and can cause serious health problems if swallowed. While many brass players may claim one particular brand of valve oil to be better than others, valve oil is no more than a petroleum distillate, with properties very similar to kerosene. In 2009, the current UK price (for 2 oz bottle) is equivalent to about £384 per gallon!

More recently, synthetic valve oils have become more readily available. Characteristics of synthetic valve oil include (but are not necessarily limited to) greater compatibility with other related oils without chemical reactions (some types of mineral based oils or their additives were known to react with each other, forming thick solids - this required disassembly and complete cleaning of the valves to restore operation), slow evaporation or total lack of evaporation resulting in fewer oilings required, and less dissipation on contact with moisture inside the valve. Some synthetics have the additional advantage that they do not act as thinners for mineral based slide greases.

Victaulic

Victaulic is a developer and producer of mechanical pipe joining systems and is the originator of the grooved coupling pipe joining system. Victaulic specialises in the development of coupling, valve, and fitting technologies for Heavy Industrial Industries such as Oil & Gas, Mining, Martime, Power, & Clean Water. Victaulic also designs couplings, valves, fittings, and sprinkler system technologies for HVAC, Plumbing, & Fire Protection in Commercial structures.

Victaulic was founded in 1925 in New York City to market a radical new concept in the piping industry — a mechanical bolted coupling that would engage into grooves and use a gasket seal. The concept of joining pipe with bolted mechanical grooved couplings originated during World War I for rapid deployment of fuel and water lines to Allied forces. The company that worked closely with the War Department of England to develop this new method of joining pipe was named Victory Pipe Joint Company after the coupling it had engineered — known as the "victory joint" from early military uses. Later, combining the words "Victory" and "Hydraulics" the name was shortened to Victaulic, and the company began to shift its focus to commercial piping applications to market its new innovative method of joining pipe. Today the company is headquartered in Easton, Pennsylvania, and employs more than 3,500 people worldwide.

Victaulic main products are:

- Victaulic Grooved Couplings used to join mechanical pipes together. The company produces a variety of these grooved couplings along with other technologies since it was founded.
- The "Victaulic Vortex" Fire Suppression System, a product placed at critical locations like data centres. It emits a mix of water and nitrogen from a single source that simultaneously cools the affected area and extinguishes the fire. The Vortex is the only Fire Suppression system approved for use in areas containing flammable liquids & hazardous machinery.
- Victaulic Sprinker Heads / Sprinkler Systems, Grooved Valves / Groove Fittings, and Expansion Joints.

Chapter 6

Vieussens Valve of the Coronary Sinus

The Vieussens valve of the coronary sinus is a valve and anatomic landmark between the coronary sinus and the great cardiac vein. It is often a flimsy valve composed of one to three leaflets found in 80% to 90% of people. It is clinically important because it is often an obstruction to catheters in 20% of patients.

Shutter (Photography)

Camera Shutter

Camera shutters can be fitted in two positions:

- Central shutters are mounted within a lens assembly, or more rarely behind or even in front of a lens, and shut off the beam of light where it is narrow. A leaf mechanism is usually used.
- Focal-plane shutters are mounted near the focal plane and move to uncover the film or sensor.

Shutters immediately behind the lens were used in some cameras with limited lens interchangeability. Shutters in front of the lens were used in the early days of photography.

Other mechanisms than the dilating aperture and the sliding curtains have been used; anything which exposes the film to light (for a specified time) will suffice.

The time for which a shutter remains open (exposure time) is determined by a timing mechanism. These were originally mechanical, but since the late twentieth century are mostly electronic.

The exposure time and the effective aperture of the lens must together be such as to allow the right amount of light to reach the film or sensor. Additionally, the exposure time must be suitable to handle any motion of the subject. Usually it must be fast enough to "freeze" rapid motion; sometimes a controlled degree of blur is desired, to give a sensation of movement.

Most shutters generate a flash synchronization switch to trigger a flash, if connected. This was quite a complicated matter with mechanical shutters and flashbulbs which took an appreciable time to reach full brightness, focal-plane shutters making this even more difficult.

Special flashbulbs were designed in a slow-burn style where the light would reach full intensity, and then remain at that intensity to wait for the slow focal-plane shutter to expose the full film frame. Many of these problems have been alleviated with modern electronic timers and electronic flash units which fire virtually instantaneously.

When using a focal-plane shutter with a flash, a photographer will typically operate the shutter at its X-sync speed or slower; however, some electronic flashes can produce a steady pulse compatible with a focal-plane shutter operated at much higher shutter speeds. Keeping in mind that the focal-plane shutter is still going to impart focal-plane shutter distortions to anything that is moving.

Cinematography uses a rotary disc shutter in movie cameras, a continuously spinning disc which conceals the image with a reflex mirror during the intermittent motion between frame exposure. The disc then spins to an open section that exposes the next frame of film while it is held by the registration pin.

Focal-Plane Shutter

In camera design, a focal-plane shutter is a type of photographic shutter that is positioned immediately in front of the focal plane of the camera, that is, right in front of the photographic film or image sensor.

Two-Curtain Shutters

The traditional type of focal-plane shutter in 35 mm cameras, pioneered by Leitz, the camera company that makes the Leica, uses two shutter curtains, made of opaque rubberized fabric, that run horizontally across the film plane. For slower shutter speeds, the first curtain opens (usually) from right to left, and after the required time with the shutter open, the second curtain closes the aperture in the same direction. When the shutter is cocked again the shutter curtains are moved back to their starting positions, ready to be released.

This is a graphical representation only; the actual mechanisms are much more complex. For example, the shutter curtains actually roll on and off spools at either side of the frame aperture so as to use as little space as possible.

Faster shutter speeds are achieved by the second curtain closing before the first one has fully opened; this results in a vertical slit that travels horizontally across the film. Faster shutter speeds simply require a narrower slit, as the speed of travel of the shutter curtains is not normally varied. Using this technique, modern SLR cameras are capable of shutter speeds of up to 1/2000, 1/4000 or even 1/8000 of a second.

The black rectangle represents the frame aperture through which the exposure is made. It is currently covered by the first shutter curtain, shown in red. The second shutter curtain shown in green is on the right side. The first shutter curtain begins to move to the left allowing the exposure to be made. Because the exposure requires a very fast shutter speed, the second curtain begins to move across at a set distance from the first one.

The first shutter curtain continues to travel across the frame aperture followed by the second curtain. It would be pointless to use an electronic flash with this shutter speed as the short duration flash would only expose a very small amount of the frame as the rest is covered by either the first or second shutter curtain.

The first shutter curtain finishes moving, followed closely by the second curtain which is now covering the frame aperture completely. When the shutter is recocked both shutter curtains are wound back to the right hand side ready for the next exposure.

Vertical-travel Shutters

Most modern 35 mm and digital SLR cameras now use vertical travel metal blade shutters. These work in precisely the same way as the horizontal shutters, but because of the shorter distance the shutter blades must travel (24 mm as opposed to 36 mm), the shutter blades can travel across the film plane in less time. This can result in faster flash synchronization speeds than are possible with the horizontal-curtain focal-plane shutter, and the shutter can reliably provide higher speeds (up to 1/8000 of a second).

Advantages

One of the advantages of focal-plane shutters is that the shutter can be built into the body of a camera which accepts interchangeable lenses, eliminating the need for each lens to have a central shutter built into it. Another advantage of the focal-plane shutter is that their fastest speeds are quite high: 1/4000 second or even 1/8000 second; much higher than the 1/500 second of the typical leaf shutter.

Disadvantages

The main disadvantage of the focal-plane shutter is that a durable and reliable one is a complex (and often expensive) device. While the concept of a travelling slit shutter is simple, a modern FP shutter is a computerized microsecond accurate timer, governing sub-gram masses of exotic materials, subjected to hundreds of g's acceleration, moving with micron precision, choreographed with other camera systems for 100,000+ cycles. This is why FP shutters are seldom seen in cheap or low quality cameras.

In addition, the typical focal-plane shutter has flash synchronization speeds that are slower than the typical leaf shutter's 1/500 sec., because the first curtain has to open fully and the second curtain must not start to close until the flash has fired. In other words, the very narrow slits of fast speeds will not be properly flash exposed. The fastest X-sync speed on a 35 mm camera is traditionally 1/60 sec. for horizontal Leica-type FP shutters and 1/125 sec. for vertical Square-type FP shutters. Modern FP shutters have increased X-sync to 1/250 sec. with the use of exotic ultra-strong materials and computer control, and 1/8000 sec. through electronic sleight of hand.

Focal-plane shutters may also produce image distortion of very fast moving objects or when panned rapidly. A large relative difference between a slow wipe speed and a narrow curtain slit results in cartoonish distortion, because one side of the frame is exposed at a noticeably later instant than the other and the object's interim movement is imaged.

For a horizontal Leica-type FP shutter, the image is stretched if the object moves in the same direction as the shutter curtains, and compressed if travelling in the opposite direction of them. For a downward firing vertical Square-type FP shutter, the top of the image leans forward. In fact, the use of leaning to give the impression of speed in illustration is a caricature of the distortion caused by the slow wiping vertical FP shutters of large format cameras from the first half of the twentieth century. Electro-optical shutters

Instead of using relatively slow-moving mechanical shutter curtains, electro-optic devices such as Pockels cells can be employed as shutters. While not commonly used, they completely avoid the problems associated with travelling-curtain shutters such as flash synchronisation limitations and image distortions when the object is moving. Such shutters are significantly more costly than mechanical shutters.

The Rotary Focal-plane Shutter

Besides the horizontal Leica and vertical Square FP shutters, other types of FP shutters exist. The most prominent is the rotary or sector FP shutter. The rotary disc shutter is common in film movie cameras, but rare in still cameras. These spin a round metal plate with a sector cutout in front of the film. In theory, rotary shutters can control their speeds by narrowing or widening the sector cutout (by using two overlapping plates and varying the overlap) and/or by spinning the plate faster or slower. For movie cameras, this is normally irrelevant, as they usually have only one speed – 1/50 sec. However, for simplicity's sake, most still camera rotary shutters have fixed cutouts and vary the spinning speed. The Olympus Pen F and Pen FT (1963 and 1966, both from Japan) half frame 35 mm SLRs spun a semicircular titanium plate to 1/500 sec.

Semicircular rotary shutters also have the advantage of unlimited X-sync speed, but all rotary FP shutters have the disadvantage of the bulk required for the plate spin. The Univex Mercury (1938, USA) half frame 35 mm camera had a very large dome protruding out the top of the main body to accommodate its 1/1000 sec. rotary shutter. They also produce very unusual distortion at very high speed because of the angular sweep of the exposure wipe. Bulk can be reduced by substituting blade sheaves for the plate, but then the rotary FP shutter essentially becomes a regular bladed FP shutter.

The Revolving Drum Focal-Plane Shutter

The revolving drum is an unusual FP shutter that has been used in several specialised panoramic cameras such as the Panon Widelux (1959, Japan) and KMZ Horizont (1968, Soviet Union). Instead of using an extremely short focal length (wide angle) lens to achieve an extra wide field of view, these cameras have a medium-wide lens encapsulated in a drum with a rear vertical slit. As the entire drum is horizontally pivoted on the lens's nodal point, the slit wipes an extra wide aspect image onto film held against a curved focal plane. The Widelux produced a 140° wide image in a 24×59 mm frame on 135 film with a Lux 26mm f/2.8 lens and controlled shutter speed by varying rotation speed on a fixed slit width.

In the Kodak Cirkut (1907, USA) and Globus Globuscope (1981, USA) cameras, the entire camera and lens revolved as the film was pulled past the slit in the opposite direction. The Globuscope produced a 360° angle of view image in a 24×160 mm frame on 135 film with a 25 mm lens and had adjustable slit width with a constant rotation speed. Revolving FP shutters produce images with unusual distortion where the image centre seems to bulge toward the viewer, while the periphery appears to curve away, because the lens's field of view changes as it swivels. This distortion will disappear if the photograph is mounted on a circularly curved support and viewed with the eye at the centre. Revolving shutters must also rotate smoothly; otherwise uneven exposure will result in ugly vertical banding in the image. Since the rotation may take several seconds to complete, no matter the shutter speed, the camera should be tripod mounted. For the same reason, flash cannot be used with these cameras.

History and Technical Development

The earliest daguerreotype (invented 1839) photographic cameras did not have shutters, because the lack of sensitivity of the process and the small apertures of available lenses meant that exposure times were measured in many minutes. A photographer could easily control exposure time by removing and returning the camera lens', lens cap or plug.

However, as one increased sensitivity process replaced another and larger apertures lenses became available during the 1800s, exposure times shortened to seconds and then to fractions of seconds. Exposure timing control mechanisms became a necessary accessory and then a standard camera feature.

The Single-curtain Focal-plane Shutter

The earliest manufactured shutter was the drop shutter of the 1870s. This was an accessory guillotine-like device – a wooden panel with slit cutout mounted on rails in front of the camera lens that gravity dropped at a controlled rate. As the slit passed the lens, it "wiped" the exposure onto the photographic plate. With rubber bands to increase the drop speed, a 1/500 or 1/1000 sec. shutter speed could be reached. Eadweard Muybridge used shutters of this type in his famous trotting horse studies.

By the 1880s, lens front mounted accessory shutter boxes were available, containing a rubberized silk cloth curtain (also called a blind) with one or more width slit cutouts wound around two parallel drums and using springs to pull a slit from one drum to the other. These shutters offered a wide range of shutter speeds by adjusting the spring tension and selecting a slit width.

In 1883, Ottomar Anschütz (Germany) patented a camera with an internal roller blind shutter mechanism, just in front of the photographic plate. Thus the focal-plane shutter in modern recognisable form was born. Goerz manufactured the Anschütz Camera (Germany) as the first production FP shutter camera in 1890. Francis Blake invented a type of focal plane shutter camera by 1889 that achieved shutter speeds of 1/2000 second, and exhibited numerous stop-action photographs. Note that a drop shutter-like mechanism with adjustable slit was used at the focal plane of an apparently one-off William England camera in 1861 and this is considered the first FP shutter of any kind.

Single curtain, vertical travelling, fixed-width slit, focal-plane shutters with adjustable spring tension and slit-width selection remained popular in large and medium format cameras for the next half-century. The lens on a single curtain FP shutter camera must have its lens cap on when the shutter is cocked; otherwise the film will be double exposed when the blind's cutout re-passes the film gate. A camera-mounted FP shutter's main advantage over the competing interlens leaf shutter was ability to use a very narrow slit to offer an action stopping 1/1000 second shutter speed at a time when leaf shutters topped out at 1/250 sec. – although the available contemporaneous ISO 1 to 3 equivalent speed emulsions limited the opportunities to use the high speeds.

However, these older focal-plane shutters wiped the exposure fairly slowly, even under the highest available spring tension, because the delicate curtain was too fragile to survive the necessary accelerative shocks to move faster. The large relative difference between a slow downward wipe speed and a narrow curtain slit resulted in cartoonish distortion of very fast moving objects instead of truly freezing their motion.

Folmer and Schwing (USA) was the most famous proponent of single curtain FP shutters, with their large format sheet film Graflex single-lens reflex and Graphic press cameras using them from 1905 to 1973. Their most common 4×5 inch shutters had four slit widths ranging from 1½ to [! inch and up to six spring tensions for a speed range of 1/10 to 1/1000 second.

The Leica-type Dual-curtain Focal-plane Shutter

In 1925, the Leica A (Germany) 35 mm camera was introduced with a dual-cloth-curtain, horizontal-travelling-slit, focal-plane shutter. A dual curtain FP shutter does not have precut slits and the spring tension is not adjustable. The exposure slit is formed by drawing open the first curtain onto one drum and then pulling closed the second curtain off a second drum after a clockwork escapement timed delay (imagine two overlapping window shades) and moves at one speed (technically the curtains are still accelerating slightly) across film gate. Faster shutter speeds are provided by timing the second shutter curtain to close sooner after the first curtain opens and narrowing the slit wiping the film. Dual curtain FP shutters are self-capping; the curtains are designed to overlap as the shutter is cocked to prevent double exposure.

Although self-capping dual curtain FP shutters date back to the late nineteenth century, the Leica design made them popular and virtually all FP shutters introduced since 1925 are dual curtain models. As perfected in the 1954 Leica M3 (West Germany), a typical Leica-type horizontal FP shutter for 35 mm cameras is pre-tensioned to traverse the 36 millimetre wide film gate in 18 milliseconds (at 2 metres per second) and supports slit widths for a speed range of 1 to 1/1000 sec. A minimum 2 mm wide slit produces a maximum 1/1000 sec. effective shutter speed. Note that the dual curtain FP shutter suffers the same fast speed distortion problems as the single curtain type. Similar technology FP shutters were also common in medium format 120 roll film cameras.

Horizontal cloth FP shutters are normally limited to 1/1000 sec. maximum speed because of the difficulties in precisely timing extremely narrow slits and the unacceptable distortion resulting from a relatively slow wipe speed. Their maximum flash synchronization speed is also limited because the slit is only fully open to the film gate (36 mm wide or wider) and able to be flash exposed down to 1/60 sec. X-synchronization (nominal; 18 ms = 1/55 sec. actual maximum; in reality, a 40 mm slit [to allow for variance] gives 1/50 sec. [S! stop slow]).

Some horizontal FP shutters managed to exceed these limits by narrowing the slit or increasing curtain velocity beyond the norm. However, these tended to be sophisticated ultra-high precision models used in expensive professional-level cameras. For example, the Nikon F2 (Japan) 35 mm single-lens reflex (SLR) camera of 1971 had a horizontal FP shutter with stronger titanium foil curtains to safely traverse the film gate in 10 ms (at 3.6 m/s and allowing 1/80 sec. X-sync) with a 1.8 mm slit giving a 1/2000 sec. top speed.

The Square-type Metal-bladed Focal-plane Shutter

In 1960, the Konica F (Japan) 35 mm SLR began a long term incremental increase in maximum shutter speeds with its “High

Synchro" FP shutter. This shutter greatly improved efficiency over the typical Leica shutter by using stronger metal blade sheaves that were "fanned" much faster, vertically along the minor axis of the 24×36 mm frame. As perfected in 1965 by Copal, the Copal Square's slit traversed the 24 mm high film gate in 7 ms (3.4 m/s). This doubled flash X-sync speed to 1/125 sec. In addition, a minimum 1.7 mm wide slit would double top shutter speed to a maximum 1/2000 sec. Note, most Squares were derated to 1/1000. sec in the interest of reliability.

The Square's metal blades were also immune to the drying out, rotting and pinholing that cloth curtained shutters could suffer from as they aged. In addition, Squares came from the supplier as complete drop-in modules, so camera designers could concentrate on camera design and leave shutter design to specialist subcontractors. This had previously been an advantage of leaf shutters.

Square-type FP shutters were originally bulky in size and noisy in operation, limiting their popularity in the 1960s among camera designers and photographers. Although Konica and Nikkormat were major users of the Copal Square, many other brands including Asahi Pentax, Canon, Leica and Minolta continued to refine the Leica-type shutter for reliability, if not speed; moving from three axis to four axis designs (one control axis for each curtain drum axis, instead of one control for both drums). New compact and quieter Square designs, with simpler construction and greater reliability, were introduced in the 1970s. The most notable were the Copal Compact Shutter (CCS), introduced by the Konica Autoreflex TC (1976), and the Seiko Metal Focal-Plane Compact (MFC), first used in the Pentax ME (1977; all from Japan). The vertical blade type supplanted the horizontal cloth type as the dominant FP shutter type in the 1980s. Even Leica Camera (originally E. Leitz), long a champion of the horizontal cloth FP shutter for its quietness, switched to a vertical metal FP shutter in 2006 for its first digital rangefinder (RF) camera, the Leica M8 (Germany).

Note that the Contax (Germany) 35 mm RF camera of 1932 had a vertical travel FP shutter with dual brass-slatted roller blinds with adjustable spring tension and slit width, and a top speed of 1/1000 sec. (the Contax II of 1936 had a claimed 1/1250 sec. top speed.), but it was woefully unreliable and not an antecedent of the modern Square shutter.

The Quest for Higher Speed

Although the Square shutter improved the FP shutter in most

ways, it still limited maximum flash X-sync speed to 1/125 sec. (Unless using special long-burn FP flash bulbs that burn throughout the slit wipe, making slit width irrelevant.) Any quality leaf shutter of the 1960s could achieve at least 1/500 sec. flash sync. Greater FP shutter X-sync speed would require further strengthening the curtains, by using exotic materials, allowing them to move even faster and widen the slits.

Copal collaborated with Nippon Kogaku to improve the Compact Square shutter for the Nikon FM2 (Japan) of 1982 by using honeycomb pattern etched titanium foil, stronger and lighter than plain stainless steel, for its blade sheaves. This permitted cutting shutter-curtain travel time by nearly half to 3.6 ms (at 6.7 m/s) and allowed 1/200 sec. flash X-sync speed. A bonus was a distortionless top speed of 1/4000 sec. (with 1.7 mm slit). The Nikon FE2 (Japan), with an improved version of this shutter, had a 3.3 ms (at 7.3 m/s) curtain travel time and boosted X-sync speed to 1/250 sec. in 1983. The top speed remained 1/4000 sec. (with 1.8 mm slit).

The fastest focal-plane shutter ever used in a film camera was the 1.8 ms curtain travel time (at 13.3 m/s) duralumin and carbon fibre bladed one introduced by the Minolta Maxxum 9xi (named Dynax 9xi in Europe, á-9xi in Japan) in 1992. It provided a maximum 1/12,000 sec. (with 1.1 mm slit) and 1/300 sec. X-sync. A further improved version of this shutter, spec'ed for 100.000 actuations, was used in the Minolta Maxxum 9 (de) (named Dynax 9 in Europe, á-9 in Japan) in 1998 and Minolta Maxxum 9Ti (named Dynax 9Ti in Europe, á-9Ti in Japan) in 1999.

The Electronically Controlled Focal-Plane Shutter

A parallel development to faster speed FP shutters was electronic shutter control as part of the general trend of electronic governance of all camera systems. In 1966, the VEB Pentacon (East Germany) was the first SLR with an electronically controlled FP shutter. It used electronic circuitry to time its shutter instead of the traditional spring/gear/lever clockwork mechanisms. In 1971, the Asahi Pentax Electro Spotmatic (Japan; name shortened to Asahi Pentax ES in 1972; called Honeywell Pentax ES in USA) tied its electronically controlled shutter to its exposure control light metre to provide electronic aperture-priority autoexposure.

The traditional 1/1000 sec. and 1/2000 sec. top speeds of horizontal and vertical FP shutters are on the knife's edge of mechanical controllability – often ¼ stop too slow, even in ultra-high-quality models. Spring powered geartrains become inadequate to durably control and reliably time any higher accelerations and shocks. For example, some highly tensioned FP shutters could suffer from "shutter curtain bounce."

This phenomenon is exactly what it sounds like – if the curtains are not properly braked after crossing the film gate, they might crash and bounce; reopening the shutter and causing double exposure ghosting bands on the image edge. Even the Nikon F2's ultra-high precision shutter suffered from this as an early production teething problem. As the Square-type FP shutter's blades moved faster and faster to provide shorter and shorter shutter speeds, the need for better blade timing control only increased.

At first, electromagnets controlled by analogue resistor/capacitor timers were used to govern the release of the second shutter curtain (though still operated by spring power). In 1979, the Yashica Contax 139 Quartz (Japan) introduced more precise digital piezoelectric quartz (shortly followed by ceramic) oscillator circuits (ultimately under digital microprocessor control) to time and sequence its entire exposure cycle – including its vertical FP shutter. Electric "coreless" micromotors, with near instantaneous on/off capability and very high power for their size, would drive both curtains (and other camera systems), replacing springs altogether, in the late 1980s. Minimising mechanical moving parts also helped to prevent inertial shock vibration problems.

Electronic control also made it simpler to time very long shutter speeds. A spring-wound clockwork escapement must completely unwind fairly quickly and limit the longest speed – generally to one full second, although the Kine Exakta (Germany) offered 12 sec. in 1936. The Olympus OM-2's electronically timed horizontal FP shutter could reach 60 sec. in 1975 and the Olympus OM-4 (both Japan) reached 240 sec. in 1983.

The Pentax LX (Japan, 1980) and Canon New F-1 (Japan, 1981) even had hybrid electromechanical FP shutters that timed their fast speeds mechanically, but used electronics to only extend the slow speed range; the LX to 125 sec., the F-1N to a modest 8 sec. Note, the Nikon F4 (Japan, 1989) was specified to reach a timed shutter speed of 999 hours with the use of the accessory electronic Multi Control Back MF-23. In theory, the longest available speed is limited only by available battery power for the electronics. This caught some 1970s photographers by surprise when they attempted very long "B" exposures and found their camera batteries had died in the middle because of the era's power hungry electronics and ruined the exposure.

Breaking the X-sync Barrier

Electronics are also responsible for pushing the focal-plane shutter's X-sync speed beyond its mechanical limits. As stated earlier, a horizontal FP shutter for 35 mm cameras is only fully open and usable for flash exposure down to 1/60 sec., while vertical FP shutters are usually limited to 1/125 sec. At higher speeds, a normal 1 millisecond electronic flash burst would only partially expose the film – the part open to the slit.

In 1986, the Olympus OM-4T (Japan) introduced a system that could synchronize a specially dedicated accessory Olympus F280 Full Synchro electronic flash to pulse its light at a 20 kilohertz rate for up to 40 ms, to illuminate its horizontal FP shutter's slit as it crossed the entire film gate – in effect, simulating long-burn FP flashbulbs – allowing flash exposure at shutter speeds as fast as 1/2000 sec. This allowed daylight plus fill-flash use in almost any situation. However, there is a concomitant loss of flash range. Extended "FP flash" sync speeds began appearing in many high-end 35 mm SLRs in the mid-1990s, and reached 1/12,000 sec. in the Minolta Maxxum 9 (de) (Japan; called Dynax 9 in Europe, Alpha 9 in Japan) of 1998. They are still offered in some digital-SLRs to 1/8000 sec. Leaf shutter cameras are not affected by this issue – they have completely different limitations.

The Focal-plane Shutter Today

Focal-plane shutter top speed peaked at 1/16,000 sec. (and 1/500 sec. X-sync) in 1999 with the Nikon D1 digital SLR. The D1 used electronic assist from its sensor for the 1/16,000 sec. speed and its 15.6×23.7 mm "APS-size" sensor was smaller than 35 mm film and therefore easier to cross quickly for 1/500 sec. X-sync.

However, with very limited need for such extremely fast speeds, FP shutters retreated to 1/8000 sec. in 2003 (and 1/250 sec. X-sync in 2006) – even in professional level cameras. In addition, since no specialised timers are needed for extremely slow speeds, the slowest speed setting is usually 30 sec.

Instead, over the last twenty years, most effort has gone into improving durability and reliability. Whereas the best mechanically controlled shutters were rated for 150,000 cycles and had an accuracy of ±¼ stop from nominal value (more typically 50,000 cycles at ±½ stop), today's best electronically controlled FP shutters can last 300,000 cycles and have no noticeable speed error.

In last few years, digital point-and-shoot cameras have been using timed electronic sampling of the image sensor, replacing the traditional mechanical leaf shutter, with delicate moving parts that can wear out, used by film P/S's. Something similar is now also occurring with more sophisticated digital cameras that, in the past, would have used focal-plane shutters. For example, the Panasonic Lumix DMC-G3 (2011, Japan) interchangeable lens digital camera has an FP shutter, but in its 20 frames per second SH Burst mode, it locks its mechanical shutter open and electronically scans its digital sensor, albeit with resolution reduced to 4 megapixels from 16 MP. In photography, a shutter is a device that allows light to pass for a determined period of time, for the purpose of exposing photographic film or a light-sensitive electronic sensor to light to capture a permanent image of a scene. A shutter can also be used to allow pulses of light to pass outwards, as in a movie projector or signal lamp.

Leaf Shutter

1. Shutter plate
2. Aperture covered by leaf shutter
3. Aperture during exposure
4. Leaf blade
5. Catch mechanism
6. Butterfly spring

A *leaf shutter* is a type of camera shutter consisting of a mechanism with one or more pivoting metal leaves which normally does not allow light through the lens onto the film, but which when triggered opens the shutter by moving the leaves to uncover the lens for the required time to make an exposure, then shuts.

Simple leaf shutters have a single leaf, or two leaves, which pivot so as to allow light through to the lens when triggered. If two leaves are used they have curved edges to create a roughly circular aperture. Simple leaf shutters typically have only one shutter speed and are commonly found in disposable cameras. Some have more than one speed. Shutters for newer digital cameras are a combination of electronic and mechanical timings. Some cameras employ a 100% electronic shutter, created by turning on and off the imaging sensor's signals. Digital cameras that can also take video implement this method for their video modes. For single-frame photography then either mechanical or mechanical+electronic methods are used.

Diaphragm Shutter

A diaphragm shutter is a type of leaf shutter consisting of a number of thin blades which briefly uncover the camera aperture to make the exposure. The blades slide over each other in a way which creates a circular aperture which enlarges as quickly as possible to uncover the whole lens, stays open for the required time, then closes in the same way. The larger the number of blades, the more accurately circular is the aperture. An odd number of blades is usually used: 3, 5, or more. The term *diaphragm shutter* has also been used to describe an optical stop with a slit, near the focal plane of a moving-film high-speed camera.

Central Shutter

A *central shutter* is a camera shutter normally located within the lens assembly where a relatively small opening allows light to cover the entire image. The term is also used for shutters behind, but near to, the lens. The alternative to a central shutter is a focal-plane shutter.

Interchangeable lens cameras with a central shutter within the lens body require that each lens has a shutter built into it. In practice most cameras with interchangeable lenses use a single focal plane shutter in the camera body for all lenses, while cameras with a fixed lens use a central shutter; many medium format cameras and most large format cameras, however, have both interchangeable lenses and use central shutters.

Film cameras, but not digital cameras, with a central shutter and interchangeable lenses often have a secondary shutter or darkslide to cover the film and allow changing lens in mid-roll without fogging the film.

Central shutters usually use either the simple leaf mechanism or the diaphragm mechanism.

The main advantages of the central shutter compared to a focal-plane shutter are:

- Relatively simple construction is possible.
- Less expensive to produce than a focal-plane shutter.
- Flash synchronization is possible at all speeds because the shutter opens fully.
- Small size is achieved by placing the shutter where the bundle of rays is smaller, either inside the lens or inside the camera body.
- Many versions have no connection between the cocking mechanism and the film advance mechanism, making multiple exposures possible.

- Generally much quieter operation because of fewer and less bulky moving parts
- More realistic photographs in high speed follow-through—lateral focal plane shutters compress or elongate the image in such cases.
- Longer shutter life.

Some disadvantages of the central shutter are:

- For an interchangeable lens system, each lens has to have a shutter built into it.
- Leaf shutter speeds are limited by the speed at which the leaves can move: normally 1/500th of a second for a diaphragm shutter and 1/125th of a second for a simple leaf shutter.
- Some versions may have no connection between the cocking mechanism and the film advance mechanism, making accidental multiple exposures a common problem, although this is a feature of camera manufacture rather than the shutter itself.

Electronic Shutter

Digital image sensors (both CMOS and CCD image sensors) can be constructed to give a shutter equivalent function by transferring many pixel cell charges at one time to a paired shaded double called frame transfer shutter. If the full-frame is transferred at one time, it is a global shutter. Often the shaded cells can independently be read, while the others are again collecting light. Extremely fast shutter operation is possible as there are no moving parts or any serialized data transfers. Global shutter can also be used for videos as a replacement for rotary disc shutters.

Image sensors without a shaded full-frame double must use time-consuming serialized data transfer of illuminated pixels called rolling shutter. It scans parts of an image mostly by line at different times causing motion effects.

Shutter Lag

Shutter lag is the time between pressing the shutter release and the camera responding by taking the picture. While this delay was insignificant on most film cameras and some digital cameras, it may be a problem when trying to capture subjects which are moving quickly such as in sports or other action photography. Today, release shutter delay might be less of an issue again due to the release lag becoming faster and faster. Release lag of a Bridge camera like the Pentax X90 is only approximately 0.021 seconds.

Shutter Cycle

A shutter cycle is the process of the shutter opening, closing, and resetting to where it is ready to open again. The life-expectancy of a mechanical shutter is often expressed as a number of shutter cycles.

Projector Shutter

In movie projection, the shutter admits light from the lamphouse to illuminate the film across to the projection screen. To avoid flicker, a double-bladed rotary disc shutter admits light two times per frame of film. There are also some models which are triple-bladed, and thus admit light three times per frame. Shutters are also used simply to regulate pulses of light, with no film being used, as in a signal lamp.

Compound Shutter

The Compound shutter is a mechanical leaf shutter for photographic cameras introduced in the late 19th century, usually used between lens components. A typical Compound shutter may have settings for T (Time), B (Bulb), and times from 1 second down to 1/100th or less.

Unlike later mechanical shutters such as the Compur which use a clockwork mechanism, shutter timing is achieved by releasing air from a small cylinder typically mounted horizontally at the top of the shutter through a small aperture. A cam turned by the shutter speed dial positions a piston in the cylinder to provide the correct exposure. When the shutter release is pressed the shutter opens and air in the cylinder is gradually released, letting the piston move; it closes the shutter when it reaches the end of its travel. The piston travels for a very short distance at the higher speeds, making their accuracy very sensitive to the precise manufacturing of the cam; lower speeds are more reliably accurate.

Kerr Cell Shutter

A Kerr cell shutter is a type of photographic shutter used for very fast shutter speeds down to nanoseconds.

A Kerr cell utilises the Kerr effect to block or transmit light.

Rotary Disc Shutter

A rotary disc shutter is a type of shutter. It is notably used in motion picture cameras.

Rotary discs are semicircular mirrors which rotate in front of the film gate, and thus expose the film. As the mirror spins it reflects the image onto the ground glass so that it can be viewed by the camera operator part of the time. The other part of the time the mirror allows the light to pass onto the film. On simple cameras this shutter is fixed and usually semi-circular. On more advanced cameras the shape of the shutter can be adjusted. This shutter setting is referred to as the shutter angle.

Adjusting the shutter angle controls the proportion of time that the film is exposed to light during each frame interval. The angle of the shutter forms a proportion to the time that each frame of film is exposed:

$$\frac{\text{Shutter Angle}}{360^{\circ}} = \frac{\text{Exposure Time}}{\text{Frame Interval}}$$

The primary reason that cinematographers adjust the shutter angle is to control the amount of motion blur that is recorded on each successive frame of film. A tight shutter angle will constrict motion blur. A wide shutter angle will allow it. A 180° shutter angle is considered normal.

So for instance, at 24p (actually 23.976 frames per second) the "Frame Interval" value is 0.04171 sec. (= 1 / 23.976). Using an "Exposure Time" of 1/50 sec., we get a "Shutter Angle" value of 173°, very close to 180° (normal motion blur effect).

This is the reason of the rule-of-thumb "Use an Exposure Time twice the value of the Frames-per-second to get normal motion blur".

Tight shutters create a stuttering stop motion animation look that has become popular in action and war films. In particular, tight shutters are used to capture particles flying through the air, such as dirt from an exploding mortar.

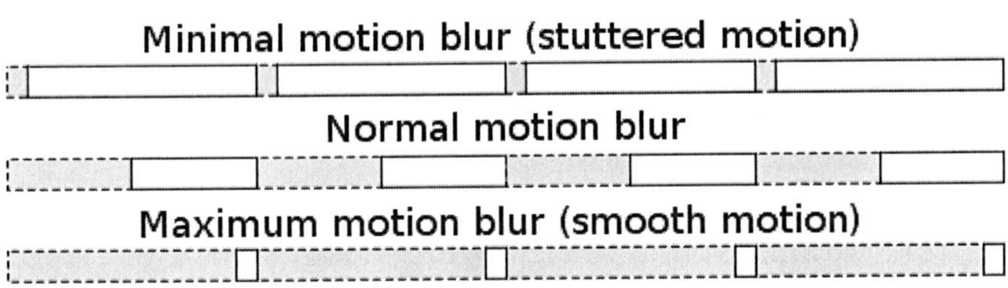

On most film cameras the shutter angle is changed by removing the lens and adjusting the shutter with a special tool. This cannot be done while the camera is operating. Some cameras such as the Arriflex 435ES can modify their shutter angle during the shot. This is sometimes referred to as an electronic shutter. An electronic shutter can compensate for the exposure change caused by a speed ramp without changing the aperture and affecting depth of field.

Other types of shutter adjustments, such as an *out-of-phase shutter*, and a *fluttering shutter* are also possible. Normally the film is held steady in the gate whenever it is exposed to light. Special shutter adjustments allow the film to be exposed while the camera mechanism is actually moving the film to the next frame. In the film *Saving Private Ryan*, cinematographer Janusz Kamiñski used such a shutter adjustment to give his film the look of World War II newsreel photography. Previously, this effect could only be achieved by purposefully mis-adjusting the timing belt of the film advance mechanism in the camera. However, several manufacturers now provide accessories to electronically control the phase relationship of the shutter and film advance mechanisms. One such device is the Timing Shift Box available for the above mentioned ARRI 435 series.

Electronic Equivalent

Many video cameras also provide the ability to adjust their shutter. In most cases the camera does not have an actual mechanical shutter. Instead, this adjustment controls the amount of time that the electronic sensor collects light in order to create each successive image. Because electronic cameras do not need to mechanically advance film, it is possible for the shutter to be open for nearly the entire period of each successive frame. At 24 frames per second, it is possible to expose a frame for nearly 1/24 second, achieving longer motion blur otherwise impossible to achieve on film at sync sound speeds.

It is even possible for the shutter to be open for multiple frames, far beyond that possible with a 360° shutter angle. One example is the Sony PD170 where the shutter can be set as low as 1/3 second; accumulating light across 10 frames (at a frame rate of approximately 30 frames of 60 interlaced fields per second for the NTSC version of the camera and 25 of 50 for the PAL version)

The term electronic shutter is often used to describe the electronic process of controlling exposure time on a light sensor. The same term is also used in film cameras to refer to a mechanical rotary shutter which can adjust its shutter angle electronically while shooting.

Rotary shutters on still cameras

While the vast majority of rotary shutters are naturally used on motion-picture cameras (as well as projectors), there were some still cameras that employed this type of shutter, notably the Univex Mercury and the Olympus Pen F SLR. Both of these cameras were half-frame 35mm cameras (frame size 18 x 24 mm). The rotary shutter proved to be very simple to construct, accurate and reliable in these cameras. Though, in the case of the Univex, it resulted in a very strange "hump" on top of the camera to cover the shutter disc.

Valvetrain

Valve cover removed on a V8 exposing valve train: rocker arms activating valve stems via pushrods.

Valvetrain is an all-encompassing term used to describe the mechanisms and parts which control the operation of the valves. A traditional reciprocating internal combustion engine uses valves to control air and fuel flow into and out of the cylinders, facilitating combustion.

Layout

Valvetrain: The valvetrain consists of valves, rocker arms, pushrods, lifters, and camshaft(s). Valvetrain opening/closing and duration, as well as the geometry of the valvetrain, controls the amount of air and fuel entering the combustion chamber at any given point in time. Timing for open/close/duration is controlled by the camshaft that is synchronized to the crankshaft by a chain, belt or gear.

Valvetrains are built in several configurations, each of which varies slightly in layout but still performs the task of opening and closing the valves at the time necessary for proper operation of the engine. These layouts are differentiated by the location of the camshaft within the engine:

Overhead Camshaft: The camshaft (or camshafts, depending on the design employed) is located above the valves within the cylinder head, and operates either indirectly or directly on the valves.

Cam-in-block: The camshaft is located within the engine block, and operates directly on the valves, or indirectly via pushrods and rocker arms. Because they often require pushrods they are often called *pushrod* engines.

Camless: This layout uses no camshafts at all. Technologies such as Solenoids are used to individually actuate the valves.

Parts

As stated above, the valvetrain is the mechanical system responsible for operation of the valves. Valves are usually of the poppet type, although many others have been developed such as sleeve, slide and rotary valves.

Poppet valves typically require small coil springs, appropriately named valve springs, to keep them closed when not actuated by the camshaft. They are attached to the valve stem ends, seating within spring retainers. Other mechanisms can be used in place of valve springs to keep the valves closed: Formula 1 engines employ pneumatic valve springs in which pneumatic pressure closes the valves, while motorcycle manufacturer Ducati uses desmodromic mechanisms to mechanically close the valves.

Depending on the design used, the valves are actuated directly by a rocker arm, finger or bucket tappet. Overhead camshaft engines use fingers or bucket tappets, upon which the cam lobes contact, while cam-in-block engines use rocker arms. Rocker arms are actuated by a pushrod, and pivot on a shaft or individual ball studs in order to actuate the valves.

Pushrods are long, slender metal rods seated within the engine block. At the bottom ends the pushrods are fitted with lifters, either solid or hydraulic, upon which the camshaft, located within the cylinder block, makes contact. The camshaft pushes on the lifter, which pushes on the pushrod, which pushes on the rocker arm, which rotates and pushes down on the valve.

Camshafts must actuate the valves at the appropriate time in the combustion cycle. In order to accomplish this the camshaft is linked to and kept in synchronisation with the crankshaft (the main shaft upon which the pistons act) through the use of a metal chain, rubber belt or geartrain. Because these mechanisms are essential to the proper timing of valve actuation they are named timing chains, timing belts and timing gears, respectively.

Desmodromic Valve

A desmodromic valve is a reciprocating engine valve that is positively closed by a cam and leverage system, rather than by a more conventional spring. The term desmodromic derives from two Greek roots, *desmos* (controlled, linked) and *dromos* (course, track).

The valves in question are those in a typical four-stroke engine that allow the air/fuel mixture into the cylinder at the beginning of the cycle and exhaust gases to be expelled at the end of the cycle. In a conventional four-stroke engine valves are opened by a cam and closed by return spring, desmodromic valves have two cams and two actuators each for positive opening and closing without any return spring.

Etymology

The word itself comes from the Greek words desmos (äåóìüò, translated as "bond" or "knot") and dromos (äñüìïò, "road" or "way"). Denoting this way the major characteristic of the valves being continuously bound to the camshaft, a tied way.

Ideal

The common valve spring system is satisfactory for traditional mass-produced engines that do not rev highly and are of a design that requires low maintenance. At the period of initial desmodromic development valve springs were a major limitation on engine performance because they would break from metal fatigue. Vacuum melt processes developed in the 1950s helped remove impurities in the steel used to make valve springs, although after sustained operation above 8000 RPM often springs would still fail. The desmodromic system was devised to remedy this problem. Furthermore, as maximum RPM increases, higher spring pressure is required to return the valve, leading to increased cam drag and higher wear on the parts at all speeds, problems addressed by the desmodromic mechanism.

Design and History

Fully controlled valve movement was conceived during the earliest days of engine development, but devising a system that worked reliably and was not overly complex took a long time. Desmodromic valve systems are first mentioned in patents in 1896 by Gustav Mees, and in 1907 the Ariès is described as having a V4 engine with "desmodromique" valve actuation, but details are scarce. The 1914 Grand Prix Delage used a desmodromic valve system (quite unlike the present day Ducati system).

Azzariti, a short lived Italian manufacturer from 1933 to 1934, produced 173 cc (173 ml) and 348 cc twin-cylinder engines, some of which had desmodromic valve gear, with the valve being closed by a separate camshaft.

The Mercedes-Benz W196 Formula One racing car of 1954-55, and the Mercedes-Benz 300SLR sports racing car of 1955 both had desmodromic valve actuation. In 1956, Fabio Taglioni, a Ducati engineer, developed a desmodromic valve system for the Ducati 125 Grand Prix, creating the Ducati 125 Desmo.

He was quoted to say: The specific purpose of the desmodromic system is to force the valves to comply with the timing diagram as consistently as possible. In this way, any lost energy is negligible, the performance curves are more uniform and dependability is better.

The engineers that came after him continued that development, and Ducati holds a number of patents relating to desmodromics. Desmodromic valve actuation has been applied to top-of-the-range production Ducati motorcycles since 1968, with the introduction of the "widecase" Mark 3 single cylinders.

In 1959, the Maserati brothers introduced one of their final designs: a desmodromic four-cylinder, 2000cc engine for their last O.S.C.A. Barchetta.

Comparison with Conventional Valvetrains

In modern engines, valve spring failure at high RPM has been mostly remedied. The main benefit of the desmodromic system is the prevention of valve float at high rpm. It has the primary disadvantages of complexity, since there are more components, and lack of understanding, which prevents people from straying from the well-known conventional valvetrain with its valve springs.

In traditional sprung-valve actuation, as engine speed increases, the momentum of the valve will eventually overcome the spring's ability to close it completely before the piston reaches TDC (Top Dead Centre). This can lead to several problems. First, and most damaging, the piston collides with the valve and both are destroyed. Second, the valve does not completely return to its seat before combustion begins. This allows combustion gases to escape prematurely, leading to a reduction in cylinder pressure which causes a major decrease in engine performance. This can also overheat the valve, possibly warping it and leading to catastrophic failure. In sprung-valve engines the traditional remedy for valve float is to stiffen the springs. This increases the seat pressure of the valve (the static pressure that holds the valve closed). This is beneficial at higher engine speeds because of a reduction in the aforementioned valve float. The drawback is that the engine has to work harder to open the valve at all engine speeds. The higher spring pressure causes greater friction (hence temperature and wear) in the valvetrain.

The desmodromic system avoids this problem, because it does not have to overcome the static energy of the spring. It still needs to work against the momentum of the valve opening and closing, and that force still depends on the effective mass of the moving parts. The effective mass of a traditional valve with spring includes one-half of the valve spring mass and all of the valve spring retainer mass. However, a desmodromic system must deal with the moment-of-inertia of the two rocker arms per valve, so this advantage depends greatly on the skill of the designer. Another disadvantage is the contact point between the cams and rocker arms. It is relatively easy to use roller tappets in conventional valvetrains, although it does add considerable moving mass. In a desmodromic system the roller would be needed at one end of the rocker arm, which would greatly increase its moment-of-inertia and negate its "effective mass" advantage. So desmo systems have generally needed to deal with sliding friction between the cam and rocker arm and therefore may have greater wear. The contact points on most Ducati rocker arms are hard-chromed to lessen this wear issue. Another possible disadvantage is that it would be very difficult to incorporate hydraulic valve lash adjusters in a desmodromic system, so the valves must be periodically adjusted.

Disadvantages

Before the days when valve drive dynamics could be analysed by computer, desmodromic drive seemed to offer solutions for problems that were worsening with increasing engine speed. Famous examples of successful desmodromic engines were Mercedes-Benz W196 and Mercedes-Benz 300 SLR racing cars. Since those days, lift, velocity, acceleration, and jerk curves for cams have been modelled by computer to reveal that cam dynamics are not what they seemed. With proper analysis, valve adjustment, hydraulic tappets, push rods, rocker arms, and above all, valve float, became things of the past without desmodromic drive.

Today most automotive engines use overhead cams, driving a flat tappet to achieve the shortest, lightest weight, and most inelastic path from cam to valve, thereby avoiding elastic elements such as pushrod and rocker arm. Computers have allowed for fairly accurate acceleration modelling of valvetrain systems.

Before numerical computing methods were readily available, acceleration was only attainable by differentiating cam lift profiles twice, once for velocity and again for acceleration. This generates so much hash (noise) that the second derivative (acceleration) was uselessly inaccurate. Computers permitted integration from the jerk curve, the third derivative of lift, that is conveniently a series of contiguous straight lines whose vertices can be adjusted to give any desired lift profile.

Integration of the jerk curve produces a smooth acceleration curve while the third integral gives an essentially ideal lift curve (cam profile). With such cams, that mostly do not look like the ones "artists" formerly designed, valve noise (lift-off) went away and valve train elasticity came under scrutiny.

Today's cams have mirror image (symmetric) profiles with identical positive and negative acceleration while opening and closing valves. An asymmetric cam either opens or closes valves more slowly than it could, speed being limited by Hertzian contact stress between curved cam and flat tappet from accelerating the mass of valve, tappet and spring.

In contrast, desmodromic drive uses two cams per valve, each with separate rocker arm (lever tappets). Maximum valve acceleration being limited by cam-to-tappet galling stress, is governed by moving mass and cam contact area. Rigidity and contact stress are best achieved with conventional flat tappets and springs whose lift and closure stress is unaffected by spring force, both occurring at the base circle where spring load is minimum and contact radius is largest. Curved (lever) tappets of desmodromic cams cause higher contact stress than flat tappets for the same lift profile, thereby limiting rate of lift and closure.

With conventional cams, stress is highest at full lift, when turning at zero speed (engine cranking), and diminishes with increasing speed as inertial force of the valve counters spring pressure, while a desmodromic cam has essentially no load at zero speed (in the absence of springs), its load being entirely inertial, and therefore increasing with speed. Its greatest inertial stress bears on its smallest radius. Acceleration forces for either method increase with the square of velocity resulting from kinetic energy.

Valve float was analysed and found to be caused largely by resonance in valve springs that generated oscillating compression waves among coils, much like a Slinky. High speed photography showed that at specific resonant speeds, valve springs were no longer making contact at one or both ends, leaving the valve floating before crashing into the cam on closure.

For this reason, today as many as three concentric valve springs are sometimes nested inside one other; not for more force (the inner ones having no significant spring constant), but to act as snubbers to reduce oscillations in the outer spring.

An early solution to oscillating spring mass was the mousetrap or hairpin spring used on Norton Manx engines. These avoided resonance but were ungainly to locate inside cylinder heads.

Valve springs that do not resonate are progressive, wound with varying pitch or varying diametre called beehive springs from their shape. The number of active coils in these springs varies during the stroke, the more closely wound coils being on the static end, becoming inactive as the spring compresses or as in the beehive spring, where the small diametre coils at the top are stiffer. Both mechanisms reduce resonance because spring force and its moving mass vary with stroke. This advance in spring design removed valve float, the initial impetus for desmodromic valve drive.

Controversy

While the desmodromic system is not ideal in a practical world of mechanics, it still survives and performs without problem. While it can be more expensive to maintain than traditional spring actuated valve systems, many aftermarket precision machined components can extend the maintenance interval to that of spring actuated systems (in comparable motorcycles).

While newer, high-performance, pneumatic systems may follow more specific design and engineering specifications (computer aided) they are typically limited to race only applications (Formula 1, Moto GP, etc.). Currently, there is no method of determining longevity or extended maintenance intervals of such systems in practical, everyday, systems such as the automobile.

While the design can be noisy, it is typically overridden by road noise from tires and other engine components such as intake and exhaust noise. Though stated above the noise is "uncomfortably loud in engines with four or more cylinders", if true, this is limited (in terms of Ducati) to the MotoGP and MotoGP Race Replica bikes, which are the only current production desmodromic motors that feature four cylinders and are intended for racing. (Note that exhaust noise levels can exceed 110 dB on full race systems.)

Examples

Famous examples include the successful Mercedes-Benz W196 and Mercedes-Benz 300 SLR race cars, and most commonly, modern Ducati motorcycles.

Ducati motorcycles with desmodromic valves have won numerous races and championships, including Superbike World Championships from 1990 to 1992, 1994–96, 1998–99, 2001, 2003–04, 2006 and 2008.

Ducati's return to Grand Prix motorcycle racing was powered by a desmodromic V4 990 cc engine in the GP3 (Desmosedici) bike, which went on to claim several victories, including a one-two finish at the final 990 cc MotoGP race at Valencia, Spain in 2006. With the onset of the 800 cc era in 2007, they are generally still considered to be the most powerful engines in the sport, and have powered Casey Stoner to the 2007 MotoGP Championship and Ducati to the constructors championship with the GP7 (Desmosedici) bike.

Overhead Valve

An overhead valve (OHV) engine, also informally called pushrod engine or I-head engine, is a type of piston engine that places the camshaft within the cylinder block (usually beside and slightly above the crankshaft in a straight engine or directly above the crankshaft in the V of a V engine), and uses pushrods or *rods* to actuate rocker arms above the cylinder head to actuate the valves. Lifters or *tappets* are located in the engine block between the camshaft and pushrods. The more modern overhead camshaft (OHC) design (still literally overhead valve) avoids the use of pushrods by putting the camshaft in the cylinder head.

In 1949, Oldsmobile introduced the Rocket V8. It was the first high-compression I-head design, and is the archetype for most modern pushrod engines. General Motors is the world's largest pushrod engine producer, producing both V6 and V8 pushrod engines.

Few pushrod type engines remain in production outside of the United States market, and even American manufacturer Ford no longer offers pushrod engines in new vehicles. This is in part a result of some countries passing laws to tax engines based on displacement, because displacement is somewhat related to the emissions and fuel efficiency of an automobile. This has given OHC engines a regulatory advantage in those countries, which resulted in few manufacturers wanting to design both OHV and OHC engines.

However, in 2002, Chrysler introduced a new pushrod engine: a 5.7 litre Hemi engine. The new Chrysler Hemi engine presents advanced features such as variable displacement technology and has been a popular option with buyers. The Hemi was on the Ward's 10 Best Engines list for 2003 through 2007. Chrysler also produced the world's first production variable valve OHV engine with independent intake and exhaust phasing. The system is called CamInCam, and was first used in the 600 horsepower (447 kW) SRT-10 engine for the 2008 Dodge Viper.

History

In automotive engineering, an overhead valve internal combustion engine is one in which the intake and exhaust valves and ports are contained within the cylinder head.

The original overhead valve or OHV piston engine was developed by the Scottish-American David Dunbar Buick. It employs pushrod-actuated valves parallel to the pistons, and this is still in use today. This contrasts with previous designs which made use of side valves and sleeve valves. Arthur Chevrolet was awarded US Patent #1,744,526 for an Overhead Valve Engine design. This patent included an adapter that could be applied to an existing engine, thus transforming it into an Overhead Valve Engine.

Nowadays, automotive use of side-valves has virtually disappeared, and valves are almost all "overhead". However most are now driven more directly by the overhead camshaft

overhead camshaft system, and these are designated OHC instead—either single overhead camshaft (SOHC) or double overhead camshaft (DOHC).

Advantages

Overhead valve (OHV) engines have specific advantages:

- *Smaller overall packaging*: because of the camshaft's location inside the engine block, OHV engines are more compact than an overhead cam engine of comparable displacement. For example, Ford's 4.6 L OHC modular V8 is larger than the 5.0 L I-head Windsor V8 it replaced. GM's 4.6 L OHC Northstar V8 is slightly taller and wider than GM's larger displacement 5.7 to 7.0 L I-head LS V8. The Ford Ka uses the venerable Kent Crossflow OHV engine to fit under its low bonnet line.
- *Less complex drive system*: OHV engines have a less complex drive system to time the camshaft when compared with OHC engines. Most OHC engines drive the camshaft or camshafts using a timing belt, a chain or multiple chains. These systems require the use of tensioners which add some complexity to the engine. In contrast a OHV engine has the camshaft positioned just above crankshaft and can be run with a much smaller chain or even direct gear connection.

Limitations

Some specific problems that remain with overhead valve (OHV) engines:

- *Limited engine speeds or RPM*: OHV engines have more valvetrain moving parts, thus more valvetrain inertia and mass, as a result they suffer more easily from valve "float", and may exhibit a tendency for the pushrods, if improperly designed, to flex or snap at high engine speeds. Therefore, OHV engine designs cannot revolve ("rev") at engine speeds as high as OHC.

Modern OHV engines are usually limited to about 6,000 to 8,000 revolutions per minute (rpm) in production cars, and 9,000 rpm to 10,500 rpm in racing applications. In contrast, many modern DOHC engines may have rev limits from 6,000 rpm to 9,000 rpm in road car engines, and in excess of 20,000 rpm (though now limited to 18,000 rpm) in current Formula One race engines using pneumatic valve springs. High-revving pushrod engines are normally solid (mechanical) lifter designs, flat and roller. In 1969, Chevrolet offered a Corvette and a Camaro model with a solid lifter cam pushrod V8 (the ZL-1) that could rev to 8,000 rpm. The Volvo B18 and B20 engines can rev to more than 7,000 rpm with their solid lifter camshaft. However, the LS7 of the C6 Corvette Z06 is the first production hydraulic roller cam pushrod engine to have a redline of 7,100 rpm.

- *Limited cylinder head design flexibility*: overhead camshaft (OHC) engines benefit substantially from the ability to use multiple valves per cylinder, as well as much greater freedom of component placement, and intake and exhaust port geometry. Most modern OHV engines have two valves per cylinder, while many OHC engines can have three, four or even five valves per cylinder to achieve greater power. Though multi-valve OHV engines exist, their use is somewhat limited due to their complexity and is mostly restricted to low- and medium-speed diesel engines. In OHV engines, the size and shape of the intake ports as well as the position of the valves are limited by the pushrods.

1994 Mercedes/Ilmor Indianapolis 500 Engine

The Indy 500 race in Indianapolis each year bears some vestige of its original purpose as a proving ground for automobile manufacturers, in that it once gave an advantage in engine displacement to engines based on stock production engines, as distinct from out-and-out racing engines designed from scratch. One factor in identifying production engines from racing engines was the use of pushrods, rather than the overhead camshafts used on most modern racing engines; Mercedes-Benz realised before the 1994 race that they could very carefully tailor a purpose-built racing engine using pushrods to meet the requirements of the Indy rules and take advantage of the 'production based' loophole, but still design it to be a state of the art racing engine in all other ways, without any of the drawbacks of a real production-based engine. They entered this engine in 1994, and because of the higher boost pressure and larger displacement that the "loophole" allowed pushrod engines, dominated the race. After the race, the rules were changed in order to reduce the amount of boost pressure supplied by the turbocharger. This amount was still 13% higher than what was allowed for the OHC engines, the engine was also allowed to retain its considerable displacement advantage. The inability of the engine to produce competitive power output after this change caused it to become obsolete after just the one race. Mercedes-Benz knew this beforehand, deciding that the cost of engine development was worth one win at Indianapolis.

Cam-in-block

The cam-in-block valvetrain layout of piston engines is one where the camshaft is placed within the cylinder block, usually beside and slightly above the crankshaft in a straight engine or directly above the crankshaft in the V of a V engine. This contrasts with an overhead camshaft (OHC) design which places the camshafts within the cylinder head and drives the valves directly or through short rocker arms.

Placing the camshaft inside the engine block has a long history in its use in valve-in-block engines, in straight and V configurations, the Ford flathead being exemplary of the type. Pushrod overhead valve engines with the cam in the block were long used in Chevrolet and Buick straight engines from the 1930s through the mid 1950s and in various similar six-cylinder engines until the extensive employment of the V6 configuration in the 1980s.

There are three main cam-in-block designs:

- L-head, also known as *L-block*, *flathead* or *sidevalve*
- F-head
- I-head, also known as overhead valve (OHV)

Flathead Engine

A flathead engine (aka sidevalve engine, flathead, or flatty) is an internal combustion engine with valves placed in the engine block beside the piston, instead of in the cylinder head, as in an overhead valve engine. As the cylinder cross-section has the shape of an upside-down L, other names such as L-block or L-head are also common.

The sidevalve's poppet valves are usually sited on one side of the cylinder(s)). A recess in the cylinder head creates a corridor connecting the valves and the combustion chamber. The valve gear comprises a camshaft which operates the valves via simple tappets, without any further valvetrain paraphernalia (such as pushrods, rocker arms, overhead valves or overhead camshafts).

The sidevalve arrangement was once the most common across all motor industries (automotive, agricultural, marine, aviation, and others), but it has since fallen from favour in most multicylinder applications, such as automotive and aviation, having been displaced by overhead-valve designs. Flathead designs are still commonly built new for many small engine applications of one and two cylinders, such as lawnmowers, rotary tillers, and two-wheel tractors.

Advantages

The main advantages of a sidevalve engine are simplicity, reliability, cheapness, compactness, responsive low-speed power, and ability to use low-octane fuel. The absence of a complicated valvetrain allows a compact engine that is cheap to manufacture. In particular, the cylinder head may be little more than a simple metal casting, and the simple valve gear engenders reliability. These advantages explain why overhead designs began their displacement of flathead designs in higher-cost, higher-performance applications (such as aircraft, luxury cars, and sports cars); only gradually worked their way into applications such as economy cars, trucking, and commercial agriculture; and still have not extensively displaced flatheads in applications such as lawn and garden equipment or subsistence agriculture.

An important benefit of the a sidevalve design is that, if a valve should drop, only limited damage would occur, and the engine might continue to operate with a broken valve. (By comparison, an OHV or OHC engine can suffer piston damage and catastrophic engine failure when a valve drops.) The design is intrinsically not an interference engine design, and provides safety from damage at high rpm.

Because intake and exhaust gases pass through the same corridor between the block and head, and because the light valve gear allows the valves to open and close quickly, flatheads are designed with minimal valve overlap (where both intake and exhaust valves are simultaneously open). This gives better low rpm performance than with OHV or OHC engines which use valve overlap to gain high rpm.

Although a sidevalve engine can safely operate at high speed, its volumetric efficiency swiftly deteriorates, so that a high power outputs are not feasible at speed.

Disadvantages

The main disadvantages of a sidevalve engine are poor gas flow, poor combustion chamber shape, and low compression ratio, all of which result in a low power output.

The maximum compression ratio is low at about 7:1, reducing efficiency but permitting the vehicle to run on low-octane fuel.

The sidevalve configuration makes intake and exhaust gases follow a circuitous route, with low volumetric efficiency, or "poor breathing". High volumetric efficiency was less important for early cars because their engines rarely sustained extended high speeds; but designers seeking higher outputs had to abandon the sidevalve. A compromise used by Willys (Jeep), Rover and Rolls-Royce in the 1950s was the F-head (aka intake-over-exhaust configuration), in which there is one side valve and one overhead valve per cylinder.

An advance in flathead technology resulted from experimentation in the 1920s by Sir Harry Ricardo, who improved their efficiency after studying the gas-flow characteristics of sidevalve engines. The Ricardo head moved the exhaust valve farther from the centre of the cylinder than the intake valve (whereas they had previously been equidistant). Ricardo observed how the form of the inlet and exhaust tracts affected gas flow and turbulence both in the inlet stream and within the combustion chamber.

Because the exhaust follows a complicated path to leave the engine, there is a tendency for the engine to overheat. In a T-block (aka T-head), a sidevalve engine has a crossflow configuration, so exhaust gases leave on the opposite side of the cylinder from the intake valve. American LaFrance powered their fire engines with T-head engines from the 1920s to the 1950s. The Cleveland Motorcycle Company produced a T-head four-cylinder in-line motorcycle engine in the 1920s; and early Stutz engines were T heads.

History and applications

Flathead inline 4- and 6-cylinder engines were frequently used for automobiles, tractors, and other products from the beginning of the automotive industry until the 1960s, when overhead valve designs began to outnumber flathead designs in production numbers. Flathead designs are still commonly built new for applications such as lawnmowers, rotary tillers, two-wheel tractors, and other one- and two-cylinder equipment.

Some of the most famous automotive flathead engines were ones whose common nicknames reflected their flathead nature, such as the Ford flathead V8 engine and the Ford Sidevalve engine.

Due to cooling and efficiency problems, flathead engines fell out of favour in "high power" applications, such as aircraft engines, prior to World War I. However they lived on for some time in the automotive world and were used on the Jeep, for instance. Flatheads are no longer in common use for automobiles (except in some rodding and customizing circles), although they are still used for some small-engine applications like lawnmowers. Because of their design, the size of valves and the compression ratio are limited, which in turn reduces available power and economy.

A new arrival is the Belgian D-Motor LF26, a compact 2.5-litre flat-four aero-engine that has direct drive to a propeller, without any reduction gearbox. Dispensing with the complexity of an OHV valvetrain, the engine gives peak power of 80 bhp at only 2800 rpm. The resulting simplicity, lightness, compactness and reliability are ideal in an aero-engine. D-Motor is developing a 120 bhp 6-cylinder version.

Harley-Davidson motorcycle flathead engines

The flathead engine saw service in Harley-Davidson motorcycles beginning with the "Sport Model" opposed twin produced from 1919 to 1923, and continuing in 1924 with single-cylinder export-model 21 cubic inches (340 cc) and 30.5 cubic inches (500 cc) singles and continued in the Servi-Cars until 1973. In the domestic U.S. market, the 45 cubic inches (740 cc) DL model (1929 to 1931) and its technical descendant, the RL model (1932 to 1936), started Harley's side-valve tradition in the 45-cubic-inch displacement class. The DL and RL models featured a total-loss oiling system and were succeeded in 1937 by the WL 45, which had recirculating oil lubrication. The WL went on to serve in WWII as the U.S. and Canadian Army's primary two-wheeled mount and subsequently as a civilian middleweight through 1952. The engine continued virtually unchanged with various G-based designations in the three-wheeled "Servi-Car" until production ceased in 1973.

In 1952, the K series flatheads was introduced, selling in parallel with the W series (which was discontinued after 1952), designed to compete with British sporting motorcycles of the time, as the American motorcycle Association allowed the 750 cc sidevalves to compete against

500 cc overhead-valve bikes. The K models featured a unit construction engine and transmission case, right side foot shift and left side foot brake, and evolved from 45 cubic inch (1952 to 1953) to 55 cubic inches by a 0.75 inches (19 mm) increase in stroke length (1954 to 1956) over its five year retail market run. The K series was replaced by the overhead valve Sportster series in the retail market in 1957. However, racing versions of the 750 cc K model, designated KR, continued to be produced in very limited numbers for some time after, winning both roadraces and dirt track events against overhead valve bikes limited to 500 cc through 1969, when the American Motorcycle Association finally decided to change the rules and make the venerable flatheads uncompetitive. The K racers were replaced first by the iron-head XR 750cc overhead valve engine, and two years later by the alloy-head XR, which continues in service in flat track racing to this day.

In 1930, the 74 cubic inches (1,210 cc) VL flathead replaced the JD Big Twin, which had featured intake-over-exhaust (IoE) valve configuration. The VL had a single downtube frame and total loss oiling, culminating in an 80 cubic inches (1,300 cc) version (VLH) in 1935. In 1937, that engine was redesigned to include a recirculating lubrication system, and designated the model U, and it went into the same frame and running gear configuration as the model E Knucklehead, which had originated in 1936. The U continued to be produced in varying configurations as a 74& cubic inch U & UL (1937 to 1948), and 80 cubic inch UH & ULH engine (1937 to 1941). By that time, the first year of the aluminium-head Panhead, it had been thoroughly superseded and outsold in the marketplace by the superior performance of the overhead valve model Big Twins.

Popular Culture

The reason that the overdrive of the Hot Rod Lincoln "just won't stall" is probably that it is a flathead V 12. (Flatheads in general, tend to have good low speed performance because of the lack of valve overlap).

T-head Engine

A T-head engine is an early type of internal combustion engine that became obsolete after World War I. It is a sidevalve engine that is distinguished from the much more common L-head by its placement of the valves. The intake valves are on one side of the engine block and the exhaust valves on the other. Seen from the end of the crankshaft, in cutaway view, the cylinder and combustion chamber resembles a T - hence the name "T-head". An L-head has all valves at the same side.

Overview

This was an early form of crossflow cylinder head. The design was very complex for its day, requiring separate camshafts to operate the intake and exhaust valves. This made the engine much more expensive to produce than a comparable L-head (flathead) engine. Additionally, it was also quite heavy and inefficient for its displacement, producing less horsepower than a flathead or modern overhead valve engine of comparable displacement.

The reason for this design's popularity from the turn of the 20th century into the 1920s is due to the type of gasoline sold at the time, which ignited at a much lower temperature than modern motor fuels. If the gasoline vapour got too hot or was compressed too much it could explode before being lit by the spark plug, a condition known as engine knocking or detonation. Since detonation was-and still is-a primary cause of catastrophic engine failure, limiting both the temperature and compression of the extremely explosive gasoline vapour was critical for maximum engine reliability. The T-Head addressed both of these issues by putting the valves in open alcoves on opposite sides of the cylinder head and having cool water from the radiator enter the engine directly over the intake valves as an extra measure of safety. The heat transfer from the exhaust ports was thus minimised, and the extra volume of the open alcoves that housed the valves also lowered the effective compression ratio of the engine. Detonation was thus eliminated, and the relative inefficiency of the engine design was offset by the dramatic reliability gains achieved and the generally lower speeds of travel and racing common to the era. This superior reliability made the design a favourite racing engine in the earliest days of auto racing, when engine reliability arguably mattered more than maximum performance. As other engine designs improved in reliability thanks to the intense engine development done in World War I, the T-head's inherent compromises in performance caused it to rapidly fall out of favour for racing applications.

Obsolescence

The T-head was rendered obsolete in the passenger car market by a combination of engine advances made in World War I and the introduction of effective anti-knock compounds such as tetra-ethyl lead in the early 1920s. When combined, these advances made flathead engines just as reliable as T-heads while also being more powerful, lighter, more fuel efficient, and cheaper to manufacture. The T-heads' incredible reliability did however keep them in service in heavy equipment, large trucks, and fire trucks until they were finally phased out of production in the 1950s.

IOE engine

The intake/inlet over exhaust (IOE) engine, also known as F-head and pocket valve, is a valvetrain configuration used in early four-stroke internal combustion engines. The configuration consists of intake valves located in the cylinder head and exhaust valves located in the cylinder or engine block.

Description

In a F-head engine, the intake manifold and its valves are located atop the cylinders, in the cylinder head, and are operated by rocker arms which reverse the motion of the pushrods so that the intake valves open downward into the combustion chamber. The exhaust manifold and its valves are located beside or as part of the cylinders, in the block. The exhaust valves are either roughly or exactly parallel with the pistons; their faces point upwards and they are not operated by pushrods, but by contact with a camshaft through the tappet or valve lifter.

Origin

The earliest IOE layouts used atmospheric inlet valves which were held closed with a weak spring and were opened by the suction created when the piston went down on the inlet stroke. When the limits of this system were reached, the design was improved without substantial changes to the head casting by adding a mechanical system to open the inlet valves and stronger springs to close them. In both cases, the exhaust valves were in the block and were opened by contact with a camshaft through a tappet or valve lifter and closed by springs.

Advantages

The IOE design allowed the use of much larger valves than a side valve or overhead valve engine. Its advantages over the flathead also included better intake mixture flow and greater reliability from its cooling of the exhaust valve and its spring.

Rover IOE Engines

Rover used a more advanced form of IOE engine. It was designed by Jack Swaine in the mid-late 1940s and was in production from 1948 to the early 1980s. Unlike the F-head, this had an efficient combustion chamber designed for good combustion, rather than simple manufacture. The top surface of the block was machined at an angle, with the piston crowns angled in a "pitched roof" to match. At TDC, the piston almost touched the angled inlet valve and provided good 'squish' to the combustion chamber itself, offset to the side by half a cylinder diametre. The resultant combustion chamber shape was a near-ideal hemisphere, although inverted and tilted from the usual "hemi-head" design. The spark plug was centrally mounted and this, together with the turbulence generated by the squish, provided a short flame path. The thinness of the gas layer between piston and inlet valve was so confined as to reduce the risk of detonation on poor fuel, one factor that kept it in service with Land Rover for so long. During the late 1940s and early 1950s when the only petrol available was low octane 'pool' petrol it also allowed Rover to run higher compression ratios than many competitors with the more usual side valve (flat head) or overhead valve designs.

The unusual combustion chamber arrangement with its angled valves also lead to an unusual valve train. The block-mounted camshaft operates small wedge shaped rockers, one for each valve. In early models the camshaft acts on a simple pad on the rocker, but for later models this pad was replaced by a roller follower. The exhaust rockers act directly on the valves, whilst the inlet rockers act on pushrods running up to a second set of longer flat rockers operating the inlet valves. The Rover engine, like many 1940s designs, was a small bore, long stroke engine to keep the RAC HP rating as low as possible, thus keeping the road tax as low as possible. The IOE layout enabled Rover to use much larger valves than would normally be possible in a small bore engine, allowing much better breathing and better performance.

The shape of the combustion chamber as an "inverted hemi-head", along with the angled cylinder head joint and pitched-roof piston crowns, had earlier been used in the 1930 Van Ranst-designed Packard V12 engine, although in this case the valves were both in the block as side valves and the spark plug was poorly placed at the extremity of the combustion chamber.

The Rover IOE engine family encompassed straight-4 (1.6- and 2.0-litres) and straight-6 (2.1-, 2.2-, 2.3-, 2.4-, 2.6- and 3.0-litres) engines and powered much of the company's post-war range in the form of the P3, P4 and P5 models. Adapted versions of the 1.6 and 2.0 IOE engines were used in early version of the Land Rover as well. Power outputs ranged from 50bhp (Land Rover 1.6) to 134bhp (P5 3 litre MkII & III). The 2.6 6-cylinder IOE engine had a particularly long career. After being used in Rover P4 saloon cars it was added to long-wheelbase Land Rover models from 1963 in the 2A Forward Control models, then in 1967 in the bonneted 109", and remained an optional fitment until 1980 when it was replaced by the Rover V8.

Other Users

The IOE valvetrain layout was used extensively in American motorcycles. Harley-Davidson used IOE engines with atmospheric inlet valves until 1912, and with mechanically driven inlet valves from 1911 to the late 1920s. Indian used IOE valvetrains on all of their four cylinder bikes except those built in 1936 and 1937. Other American motorcycle manufacturers that use IOE engines include Excelsior, Henderson, and Ace.

Hudson used an IOE straight-six engine from 1927 to 1929. Willys, and its successor Kaiser-Jeep, used variants of the Willys Hurricane engine from 1950 to 1971. Rolls-Royce used variants of their B60 IOE straight-six engine in their Wraith and Silver Wraith cars. A short-stroke oversquare version of the B60 engine was also used in the Princess 4-litre-R saloon car, displacing 3909cc and producing around 175 hp (manufacturers figures). This was known as the FB60 and was substantially upgraded from the common-or-garden B60 - it had hydraulic lifters, different carburation, hotter cam and other improvements. These changes liberated a 55 hp increase from the B60 engines' 120 hp output. Over 6000 of these cars were made.

Exhaust Over Intake (EOI)

In 1936 and 1937, the Indian Four had the valve positions reversed, with the exhaust valve in the head and the inlet valve in the block. In theory, this would improve fuel vapourisation, and the engine was actually more powerful. However, the new system made the cylinder head very hot. The exhaust valve linkage required frequent adjustment. The design was returned to the original IOE configuration in 1938.

Chapter 7

Caprotti Valve Gear

The Caprotti valve gear is a type of steam engine valve gear invented in the early 1920s by Italian architect and engineer Arturo Caprotti. It uses camshafts and poppet valves rather than the piston valves used in other valve gear. While basing his design on automotive valves, Caprotti made several significant departures from this design to adapt the valves for steam.

In the 1950s, Caprotti valve gear was improved and this British Caprotti valve gear was fitted to the last two British Railways-built 'Black Fives' 44686/7, the last 30 BR standard class 5s, numbers 73125-54 and the unique BR standard class 8 71000 Duke of Gloucester. Results were mixed, with the performance of the *Duke of Gloucester* particularly disappointing but this was later found to be due to errors elsewhere in the design and construction.

Although more expensive to manufacture than its rivals, this improved Caprotti valve gear is considerably more efficient than any other. Benefits include much of the mechanism enclosed thus leading to reduced wear and tear from the harsh steam locomotive environment and completely independent control of admission and exhaust. The restored *Duke of Gloucester*, with its flaws eliminated, has proved the concept.

Gresley Conjugated Valve Gear

The Gresley conjugated valve gear is a valve gear for steam locomotives designed by Sir Nigel Gresley, chief mechanical engineer of the LNER, assisted by Harold Holcroft. It enables a three-cylinder locomotive to operate with only the two sets of valve gear for the outside cylinders, and derives the valve motion for the inside cylinder from them by means of levers (the "2 to 1 lever" and "equal lever"). The gear is sometimes known as the Gresley-Holcroft gear, acknowledging Holcroft's major contributions to its development.

Operation

The Gresley conjugated gear is effectively an adding machine, where the position of the valve for the inside cylinder is the sum of the positions of the two outside cylinders, but reversed in direction. It can also be thought of as a rocking lever between one outside cylinder and the inside cylinder, as is common on 4-cylinder steam locomotives, but with the pivot point being moved back and forth by a lever from the other outside cylinder.

If we approximate the motion of each valve by a sine wave — if we say the position of a valve in its back-and-forth travel is exactly proportional to the sine of the "driver angle", once we have set the zero point of driver angle at the position it needs to be for that valve — then the mathematics is simple. The position of the valve that is pinned to the long end of the 2-to-1 lever is sinθ, while the positions of the other two valves are supposed to be sin(θ+120°) and sin(θ-120°). The position of the short end of the 2-to-1 lever is $-\frac{1}{2}\sin\theta$ — which, it turns out, is midway between and for any value of. So a 1-to-1 lever pivoted on the short arm of the 2-to-1 lever will do the trick.

Crank Angles

Locomotives with Gresley valve gear must have the three pistons operating at precisely 120 degree intervals. (Different spacings could be accommodated by different lever proportions, if there were any advantage to a spacing other than 120-120-120.) In order for the inside crank to clear the leading coupled axle, the inside cylinder of a locomotive with Gresley valve gear is typically positioned higher than the outside cylinders and angled downward. To maintain a smooth flow of torque, the crank angles are offset from equal 120 degree spacing to compensate for the angle of the inside cylinder (eg. 120/113/127 degrees). The resultant timing of the blast from steam exiting the cylinders still gives these three-cylinder locomotives a regular exhaust beat.

Problems

The main difficulty with this valve gear was that at high speeds, inertial forces caused the long conjugating lever to bend or "whip". This had the effect of causing the middle cylinder to operate at a longer cutoff than the outer cylinders, and therefore to produce a disproportionate share of the total power output, leading to increased wear of the middle big end. Sustained high speed running could sometimes cause the big end to wear rapidly enough that the increased travel afforded to the middle piston by the increased play in the bearing was enough to knock the ends off the middle cylinder. Although the problem could be contained in a peacetime environment with regular maintenance and inspections, it proved to be poorly suited to the rigours of heavy running and low maintenance levels of World War 2. This gave rise to big-end problems on the centre cylinder connecting rod on the famous A4 class of streamlined Pacifics and many of these locomotives were fitted with a reduced diametre piston and had the inside cylinder lined up as a temporary measure. Gresley's successor at the LNER, Edward Thompson, was critical of this particular valve gear. As well as introducing new two-cylinder designs, he set about rebuilding Gresley locomotives with separate sets of Walschaerts valve gear for each cylinder.

USA and Australia

Gresley conjugated valve gear was used by the American Locomotive Company under license and the 4-12-2 locomotives they built for the Union Pacific Railroad between 1926 and 1930 were the largest locomotives to use this valve gear. It was also used in Australia for the Victorian Railways S class 4-6-2 of 1928 and New South Wales Government Railways D57 class 4-8-2 of 1929.

As in the UK, the mechanism was not without its problems. Some of the Union Pacific 9000 class locomotives were converted to a "double Walschaerts" valve gear, while later examples were built with roller bearings for the moving parts of the Gresley mechanism. In Australia, later VR and NSWGR 3 cylinder locomotive designs used alternative mechanisms to the Gresley system in an effort to overcome its high maintenance overhead. The Victorian Railways H class of 1941 was fitted with a German Henschel and Sohn conjugated valve gear mechanism which was judged to be superior to the Gresley system, while in New South Wales the D58 class of 1950 utilised a rack and pinion system which while in theory an improvement over the Gresley system, proved in practice to be even more problematic.

Walschaerts Valve Gear

The Walschaerts valve gear is a type of steam engine valve gear invented by Belgian railway mechanical engineer Egide Walschaerts in 1844. The gear is sometimes named without the final "s", since it was incorrectly patented under that name. It was extensively used in steam locomotives from the late 19th century until the end of the steam era.

History

The Walschaerts valve gear was slow to gain popularity. The Stephenson valve gear remained the most popularly used valve gear on 19th-century locomotives. The Walschaerts gear had the advantage that it could be mounted entirely on the outside of the locomotives, leaving the space between the frames clear; this caused adoption first among some articulated locomotives.

In 1874, New Zealand Railways ordered two types of Double Fairlie locomotives from Avonside. Both the B class and E class Double Fairlies were fitted with Walschaerts valve gear. This was certainly the first use of this technology to be used in New Zealand, and is possibly the first time a British manufacturer has supplied it.

The Mason Bogie locomotive type was the first to use the Walschaerts gear in North America.

The first application in Britain was on a Single Fairlie 0-4-4T, exhibited in Paris in 1878 and purchased by the Swindon, Marlborough and Andover Railway in 1883. According to Ahrons, the locomotive saw very little service as nobody seems to have known how to set the valves and this led to enormous coal consumption.

In the 20th century, the Walschaerts valve gear was the most commonly used type, especially on larger locomotives. In Europe, its use was almost universal, whilst in North America, the Walschaerts gear outnumbered its closest competitor, the Baker valve gear, by a wide margin.

In Germany, the Walschaerts gear is often named the Heusinger valve gear after Edmund Heusinger von Waldegg, who invented the mechanism independently in 1849. Heusinger's gear was closer to the form generally adopted, but most authorities accept Walschaerts' invention as sufficiently close to the final form.

Purpose

The Walschaerts valve gear is an improvement on the earlier Stephenson valve gear in that it enables the engineer to operate the steam engine in a continuous range of settings from maximum economy to maximum power. At any setting, the valve gear satisfies the following two conditions:

- At the instant when the space on one side of the piston starts to expand, i.e. at the very start of a stroke, the valve opens to admit steam from the boiler into that space. The pressure of this steam provides the driving force.

- At the instant when the space on one side of the piston starts to contract, the valve starts to release steam from that space to the atmosphere, so as not to impede the movement of the piston.

These functions are shown in the animated diagram below.

In an economical setting, steam is admitted to the expanding space for only part of the stroke; at a point set by the engineer, the intake is cut off. Since the exhaust is also shut, during the rest of the stroke the steam that has entered the cylinder expands in isolation, and so its pressure decreases. For maximum economy, the engineer carefully sets the cutoff point so that, when the exhaust valve opens, the steam is down to exactly atmospheric pressure. Thus, all the mechanical energy available from the steam (in the absence of a condenser) is used.

The Walschaerts valve gear enables the engineer to change the cutoff point without changing the points at which intake starts and at which exhaust starts. With the Stephenson valve gear, in contrast, if the engineer shortens the intake period, the openings of the intake and exhaust are delayed, occurring after a stroke has begun.

Economy also requires that the throttle is wide open, so that no energy is wasted pushing steam through a constriction, and that the boiler pressure is at the maximum safe level to maximise thermal efficiency. For economy, a steam engine is used of a size such that the most economical settings yield the right amount of power most of the time, such as when a train is running at steady speed on level track.

When greater power is necessary, e.g. when gaining speed when pulling out of a station and when ascending a gradient, the Walschaerts valve gear enables the engineer to set the cutoff point near the end of the stroke, so that the full pressure of the boiler is exerted on the piston for almost the entire stroke. With such a setting, when the exhaust opens, the steam in the cylinder is near full boiler pressure. The pressure in the steam at that moment serves no useful purpose; it is wasted driving a sudden pulse of pressure into the atmosphere, but this waste is more than compensated by maximised economy at other times.

This sudden pulse of pressure causes the loud "choo" sound that members of the public associate with steam engines, because they mostly encounter engines at stations, where efficiency is sacrificed as trains pull away. A steam engine well adjusted for efficiency makes a soft “hhHHhh” sound that lasts throughout the exhaust stroke, with the sounds from the two cylinders overlapping to produce a nearly constant sound.

Technical Details

The heart of the Walschaerts valve gear is the combination lever. Its lower end moves in phase with the piston rod; its upper end moves 90° out of phase with the piston rod. The valve stem thus moves in a combination of these two motions. Through the reach rod and the radius rod lifter, the engineer adjusts the position of the die block in the expansion link, and thus adjusts the amplitude of the motion of the upper end of the combination lever (including the possibility of reversal of that motion, which reverses the engine).

To make possible a period of cutoff for efficiency, the valve is designed so that both intake and exhaust are blocked when the valve rod is in a range centred at the centre of its travel. This range is its lap. When maximum power is sought, the exhaust opens almost immediately after the intake closes. Therefore, the valve must move rapidly through this middle range of cutoff during such operation. Since this rapid transition occurs when the piston passes through dead centre, the motion of the valve must be 90° out of phase with that of the piston rod. The engineer can satisfy these two conditions by moving the reach rod to an extreme position, so that the die block is near an end of the expansion link, where the motion is fastest and its amplitude is greatest. The upper end of the combination lever then dominates the motion of the valve stem; the valve is in the cutoff range only briefly, and its phase is near 90°.

For more efficient operation, the engineer moves the die block near the centre of the expansion link. The amplitude of the valve stem’s motion is then lower, and its phase is shifted away from 90°. The Walschaerts valve gear, when correctly designed, provides constant valve lead irrespective of the chosen cutoff in both forward and reverse gear. With the piston in the front or rear dead centre positions, the expansion link should be vertical and moving the die block up and down should not alter the valve spindle position.

The amount of lead provided is determined by the proportions of the combination lever and the lap designed into the valve. It can only be changed by changing the proportions of these components.

The valve gear operation combines two motions; one is the lead motion (imparted at the bottom of the combination lever) and the other is the directional motion required in full gear (imparted at the top of the combination lever). To understand the operation, consider the cutoff in mid-gear. Because the die block does not move, the end of the radius rod is in effect a fixed point about which the combination lever pivots, so the valve motion is that caused directly by the cross head motion acting through the union link and combination lever. This moves the valve in each direction by a distance which should be set equal to lap plus lead of the valve, so that intake and exhaust both begin at the start of a stroke, i.e. at dead centre.

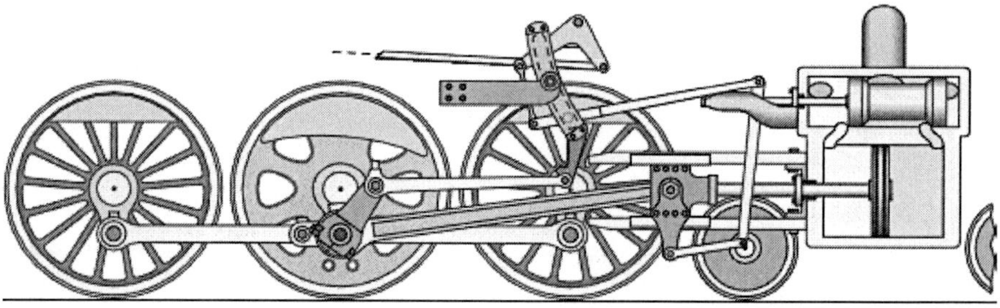

Figure: Animation of forwards motion of the gear. In this animation, the red represents live steam entering the cylinder, while the blue represents expanded (spent) steam being exhausted from the cylinder.

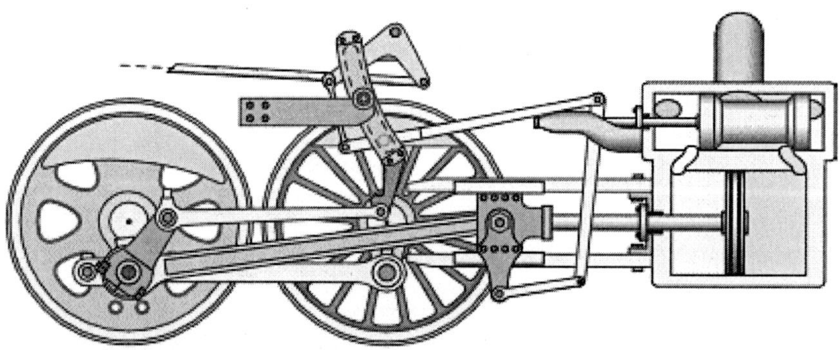

Figure: Animation showing changes occurring to reverse

If this were the only motion made by the valve then the engine would only start if the piston were close to dead centre, and even then the direction in which the engine moved would depend upon whether it was fractionally before or after the dead centre point.

Outside Admission Valves

Consider the case of outside admission valves such as slide valves. A second motion is superimposed on the crosshead motion to the valve; the return crank provides a motion which leads the crankpin motion by 90 degrees. By choice of position of the reach rod the die block may be moved from top to bottom which superimposes a varying portion of this motion ranging from either 90 degrees in lead ahead of the crank (full forward gear) through none in midgear to 90 degrees lagging the crank (full reverse gear). Lag and lead of these motions are always described relative to the engine moving forwards. The larger the magnitude of the superimposed component, the later the cutoff and thus the less expansion allowed for the steam in the cylinder.

Inside Admission Valves

For inside admission valves (most piston valves, but there are some exceptions on rebuilt locomotives) the lag and lead motions must be reversed usually by locating the eccentric crank pin 90 degrees after (following) the main crank pin (in forward motion) and the combination rod pivots on the radius rod above the linkage to the valve rods instead of below. This is the case shown in the diagrams above.

Layout

To lay out the Walschaerts gear, the proportions of the combination lever must be chosen. The total movement of the valve rod in mid gear should be symmetric between front and back dead centre. A displacement of the union link end by half the piston travel must cause a valve rod displacement of the valve lead plus the valve lap. The ratio of distance from union link end to pivot with radius rod to the distance between the valve rod end to the pivot with the radius rod should be in the same proportion as half piston travel to valve lap plus lead.

Next, the fact that in midgear the valve opening should be the same for the piston in both front and rear dead centre positions is used to find the mid gear position of the pivot between the combination rod and reach rod. In practice the valve spindle length is adjusted to ensure this as part of setting the timing. Also the dimensions of the lead and lap lever are set so that the valve moves twice the lead + lap of the valve.

With the reverser in mid gear position since there should be no superimposed return crank motion, the expansion link die slot should be an arc of a circle centred on the pivot in mid gear and of radius equal to the length of the radius rod. This condition precludes adjustment of the valve gear by altering the length of the radius rod.

The throw of the return crank must now be determined; The actual throw required from the crank is determined from the dimensions of the expansion link.

Variants

There have been many variants of Walschaerts valve gear, including:

- Baker valve gear
- Baguley valve gear
- Bagnall-Price valve gear
- J. T. Marshall valve gear

Holcroft Valve Gear

The Holcroft valve gear was a type of valve gear designed by Harold Holcroft and used on three-cylinder steam locomotives of the South Eastern and Chatham Railway (SECR). It bore many similarities to the Gresley conjugated valve gear used on the original LNER A1 design. It varied from the Gresley method of operation by using the combination lever assembly instead of the valve spindles to drive the middle cylinder of a three-cylinder design. This had operational advantages and resolved some of the problems associated with Gresley's design.

Background

When working for the GWR, Holcroft collaborated with Sir Nigel Gresley to invent and patent the Gresley conjugated valve gear for three-cylinder steam locomotives. The Gresley application of conjugated valve gear required further development to increase reliability and promote ease of maintenance. The problems associated with these issues reduced efficiency when the locomotive was under way, which made LNER locomotives sound "off-beat", and increased the incidence of damage to the track. In an effort to resolve these shortcomings, Holcroft set about designing a new, more reliable version of the Gresley conjugated valve gear.

Design

Holcroft's valve gear assembly was designed to regulate the timing of valve events admitting steam into the three cylinders. This increased efficiency whilst reducing mechanical wear and the effects of hammerblow on the track. To achieve this, Holcroft introduced a method of driving a middle cylinder from the right-hand set of "outside"

Walschaerts valve gear, eliminating the use of a separate set of valve gear between the frames.

The middle cylinder was driven from the combination lever assembly that controlled the fore and aft movement of the valve spindle on the right-hand "outside" cylinder. This admitted "live" steam and ejected "spent" steam through the ports of the middle cylinder in relation to the position of the "outside" cylinder's combination lever.

Holcroft's conjugation mechanism was a revision of Nigel Gresley's method of operating the "inside" cylinder via an extension of the "outside" valve spindles on LNER locomotives. Instead, Holcroft carried-forward the action of the combination lever assembly that controlled the fore and aft movement of the valve spindle (which admitted "live" steam into the cylinder and ejected "spent" steam out through the ports) to a pivot point connected to a lever from the "inside" valve spindle. The Holcroft variant held an advantage over Gresley's "conjugated" design, as the valve timing was immune to variations brought about by heat expansion of the valve spindles and flexing of the conjugation assembly when under heavy use.

Joy Valve Gear

Joy valve gear is a type of steam locomotive valve gear, designed by David Joy (1825–1903), Locomotive and Marine engineer, and patented in 1870. The movement is derived from a vertical link connected to the connecting rod. The vertical movement is translated into the horizontal movement required by the valve spindle by a die block moving in a slide which can be varied in inclination. The necessary 90 degree phase shift is obtained not from an eccentric, as with most designs, but from the 90 degree phase difference between the horizontal and vertical components of the motion of the connecting rod.

Applications

Joy valve-gear was used extensively on Lancashire and Yorkshire Railway and London and North Western Railway locomotives, and also on the Manning Wardle-built narrow gauge 2-6-2 locomotives of the Lynton and Barnstaple Railway. Joy gear was also used by John Fowler & Co. and several other steam traction engine manufacturers.

Preservation

The last locomotives with Joy valve gear to remain in service with British Railways were the LNWR G2 and G2A classes, one of which is preserved.

Southern Valve Gear

Southern valve gear was briefly popular on steam locomotives in the United States. It combines elements of the Walschaerts and Baker patterns.

History

Southern valve gear was devised by locomotive designers on the Southern Railway (US) and used on many locomotives on that line. The first patent was issued to Wm. Sherman Brown in 1906. He continued work and another patent was issued July 23, 1912. The gear was first tested on Mikado 586 in February, 1913 at Southern Knoxville Shops. It was specified for some USRA standard locomotive designs, though many locomotives constructed to these patterns used Walschaerts or Baker valve gear instead. The Southern valve gear was used on many Southern Railway locomotives, including survivors 630 and 722. It was also used on East Broad Top numbers 16-18, which survive today.

Operation

Unlike the Walschaerts and Baker systems, the Southern valve gear dispensed with the combination lever which transmitted some of the crosshead motion to the valves. As with those systems it used a return crank on a driver (instead of the eccentrics used by the Stephenson valve gear). The return crank attached to the eccentric rod, which hung from a pivot close to the far end of the eccentric rod. This pivot attached to the end of the radius hanger, the other end of which attached to a link block which slid back and forth in a track curved in the same manner as the expansion link of a Walschaerts gear. In this system, however, the track was fixed in place; sliding the link block back and forth controlled reversing and cutoff.

At the end of the eccentric rod, close to the radius hanger, the rod connected to the transmission yoke. The upper end of this rod connected to a bellcrank, which translated the up and down motion of the transmission yoke into back and forth motion of the valves.

The mechanism is a little less obvious than that of the other types, but control is fundamentally like that of the Baker system. In this case the controlling factor is the angle between the radius hanger and the transmission yoke; when they are parallel, there is little up-down motion of the transmission yoke and the engine is centred. As link block moves back and forth, the angle of the radius hanger changes, and the up and down motion of the transmission yoke in response to the back and forth motion of the eccentric rod is increased or decreased.

Stephenson Valve Gear

Historical Background

During the 1830s the most popular valve drive for locomotives was known as gab motion in the U.K. and V-hook motion in the U.S.A. The gab motion incorporated two sets of eccentrics and rods for each cylinder; one eccentric was set to give forward and the other backwards motion to the engine and one or the other could accordingly engage with a pin driving the distribution valve by means of the gabs: - vee-shaped ends to the eccentric rods supposed to catch the rocker driving the valve rod whatever its position. It was a clumsy mechanism difficult to operate that moreover only gave fixed valve events.

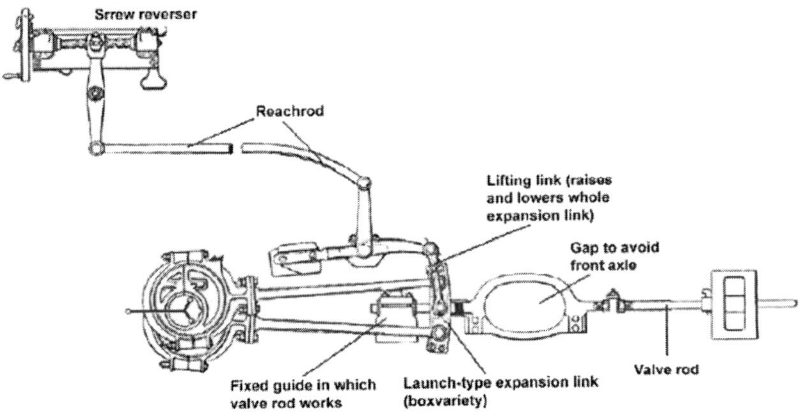

Figure: Inside Stephenson valve gear as applied to a French 0-6-0 outside cylinder mixed traffic locomotive (Midi 801) in 1867

In 1841 two employees in Stephenson's locomotive works, draughtsman William Howe and pattern-maker William Williams, suggested the simple expedient of replacing the gabs with a vertical slotted link, pivoted at both ends to the tips of the eccentric rods. To change direction, the link and rod ends were bodily raised or lowered by means of a counterbalanced bell crank worked by a *reach rod* that connected it to the reversing lever. This not only simplified reversing but it was realised that the gear could be raised or lowered in small increments, and thus the combined motion from the "forward" and "back" eccentrics in differing proportions would impart shorter travel to the valve, cutting off admission steam earlier in the stroke and using a smaller amount steam expansively in the cylinder, using its own energy rather than continuing to draw from the boiler. It became the practice to start the engine or climb gradients at long cutoff, usually about 70-80% maximum of the power stroke and to shorten the cutoff as momentum was gained in order to benefit from the economy of expansive working and the effect of increased lead and higher compression at the end of each strike. This process was popularly known as *"linking up"* or *"notching up"*, the latter because the reversing lever could be held in precise positions by means of a catch on the lever engaging notches in a quadrant; the term stuck even after the introduction of the screw reverser. A further intrinsic advantage of the Stephenson gear not found in most other types, was variable lead. Depending on how the gear was laid out, it was possible to considerably reduce compression and back pressure at the end of each piston stroke when working at low speed in full gear; once again as momentum was gained and cutoff shortened, so lead was automatically advanced and compression increased, cushioning the piston at the end of each stroke and heating the remaining trapped steam in order to avoid temperature drop in the fresh charge of incoming admission steam.

American locomotives universally employed inside Stephenson valve gear placed between the frames until around 1900 when it quickly gave way to outside Walschaerts motion. In Europe, Stephenson gear could be placed either outside the driving wheels and driven by either eccentrics or return cranks or else between the frames driven from the axle through eccentrics, as was mostly the case in Great Britain.

Applications

Abner Doble considered Stephenson valve gear: "(...) the most universally suitable valve gear of all, for it can be worked out for a long engine structure or a short one. It can be a very simple valve gear and still be very accurate, but its great advantage is that its accuracy is self-contained, for the exact relationship between its points of support (eccentrics on shaft, valve crosshead, and link hanger arm) have but little effect on the motion of the valve. Its use on engines in which all the cylinders lie in one plane, represents, in the belief of the writer, the best choice." *Another benefit of the Stephenson gear, intrinsic to the system, is variable lead: usually zero in full gear and increasing as cutoff is shortened.*

Stephenson valve gear is a convenient arrangement for any engine that needs to reverse and was widely applied to railway locomotives, traction engines, steam car engines and to stationary engines that needed to reverse, such as rolling-mill engines. It was used on the overwhelming majority of marine engines.

Details of the gear differ principally in the arrangement of the expansion link. In early locomotive practice, the eccentric rod ends were pivoted at the ends of the link while, in marine engines, the eccentric rod pivots were set behind the link slot (or below on a vertical engine).

These became known respectively as the 'locomotive link' and the 'launch link'. The launch link superseded the locomotive type as it allows more direct linear drive to the piston rod in full gear and permits a longer valve travel within a given space by reducing the size of eccentric required for a given travel. Launch-type links were pretty well universal for American locomotives right from the 1850s but, in Europe, although occurring as early as 1846, they did not become widespread until around 1900.

Larger marine engines generally used the bulkier and more expensive marine double-bar link, which has greater wearing surfaces and which improved valve events by minimising geometric compromises inherent in the launch link.

Derivatives

As a harmonic valve gear, the Stephenson arrangement may be considered as optimum. Nevertheless the fact the link needed to be bodily displaced in order to reverse meant that it required considerable vertical clearance.

At the time of its introduction, it was deemed important in the locomotive world to keep the centre of gravity, and therefore the boiler centre line as low as possible. Because valve gears in Britain were generally placed between the frames beneath the boiler, the extremely cramped conditions made the valve gear inaccessible for servicing. Also reversing could be a strenuous occupation as it entailed lifting the weight of the link plus eccentric rod ends. In order to address these problems two main variants were developed :

Gooch Valve Gear

In the Gooch valve gear the reversing and cut-off functions were achieved by raising or lowering a radius rod which connected the valve-rod to a "stationary" link pivoting around a fixed point. The advantages sought were reduced height for the gear and lighter action as the reversing lever was only required to lift the weight of the radius rod.

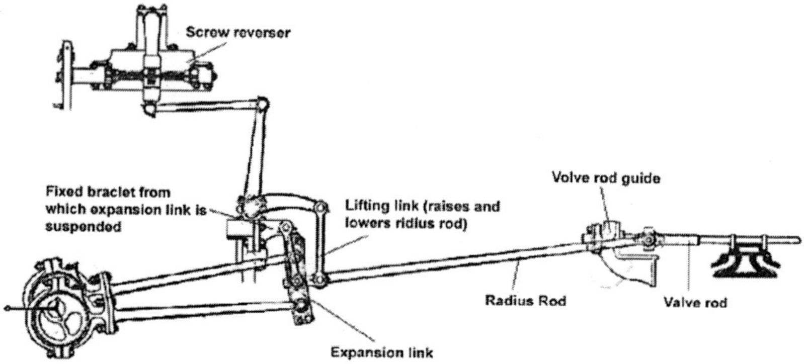

Figure: Gooch outside valve gear as applied to a French 2-4-0 outside cylinder express locomotive (Midi no. 51) in 1878

This meant that the link was convex (in relation to the eccentrics) instead of concave. Gooch valve gear had the disadvantage of angularity between the valve spindle and the eccentric rod in full gear, whereas the best forms of the Stephenson gear, the thrust was in a straight line. The Gooch gear gave constant lead at whatever cutoff. This was observed to be a disadvantage when similar locomotives fitted with either Gooch or Stephenson gear were compared in service Gooch gear was never popular in Britain except with one or two engineers down to the 1860s, but it was quite common in France.

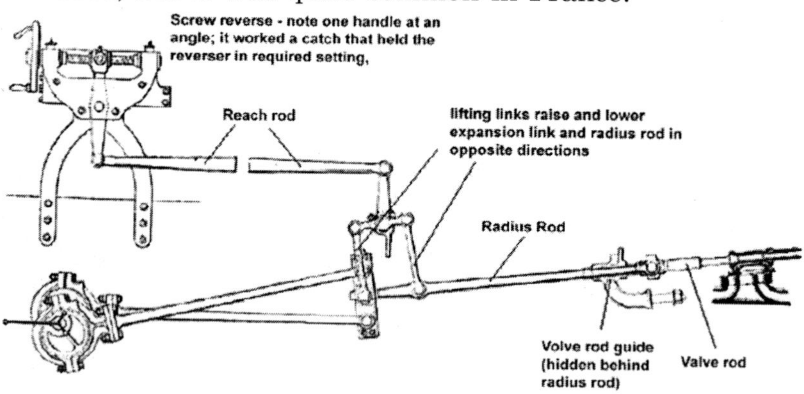

Figure: Allan outside valve gear as applied to a French 0-8-0 outside cylinder freight locomotive (Midi no.701) in 1863

Allan Straight Link Valve Gear

The Allan straight link valve gear (invented by Alexander Allan in 1855) combined the features of the Stephenson and Gooch gears. The reversing and cut-off functions were achieved by simultaneously raising the radius rod and lowering the link or vice-versa. As with the Gooch gear, this saved space but the Allen gear gave performance closer to that of the Stephenson. Moreover the straight expansion link simplified manufacture. Once again, the Allan gear was not often used in the UK but fairly common on the Continent. Notable UK examples are narrow gauge locomotives produced by George England (e.g. "Prince" & "Palmerston" in preservation on the Ffestiniog Railway) and Fletcher-Jennings (No 1 "Tallyllyn" & No 2 "Dolgoch" in preservation on the Talyllyn Railway).

Valve Gear

The valve gear of a steam engine is the mechanism that operates the inlet and exhaust valves to admit steam into the cylinder and allow exhaust steam to escape, respectively, at the correct points in the cycle. It is sometimes referred to as the "motion".

In the simple case, this can be a relatively simple task as in the internal combustion engine in which the valves always open and close at the same points. This is not the ideal arrangement for a steam engine, though, because greatest power is achieved by keeping the inlet valve open throughout the power stroke (thus having full boiler pressure, minus transmission losses, against the piston throughout the stroke) while peak efficiency is achieved by only having the inlet valve open for a short time and then letting the steam expand in the cylinder (expansive working).

The point at which steam stops being admitted to the cylinder is known as the *cutoff* and the optimal position for this varies depending on the work being done and the tradeoff desired between power and efficiency. Steam engines are fitted with regulators (throttles in US parlance) to vary the restriction on steam flow, but controlling the power via the cutoff setting is generally preferable since it makes for more efficient use of boiler steam.

A further benefit may be obtained by admitting the steam to the cylinder slightly before front or back dead centre. This advanced admission (also known as *lead steam*) assists in cushioning the inertia of the motion at high speed.

In the internal combustion engine, this task is performed by cams on a camshaft driving poppet valves, but this arrangement is not commonly used with steam engines, partly because achieving variable engine timing using cams is complicated. Instead, a system of eccentrics, cranks and levers is generally used to derive the motion of a D slide valve or piston valve from the motion of the main crank/big end. Generally, two simple harmonic motions with different fixed phase angles are added in varying proportions to provide an output motion that is variable in phase and amplitude. A variety of such mechanisms have been devised over the years, with varying success.

Both slide and piston valves have the limitation that intake and exhaust events are fixed in relation to each other and cannot be independently optimised. Lap is provided on steam edges of the valve, so that although the valve stroke reduces as cutoff is advanced, the valve is always fully opened to exhaust. However, as cutoff is shortened, the exhaust events also advance. The exhaust release point occurs earlier in the power stroke and compression earlier in the exhaust stroke.

Early release wastes some energy in the steam, and early closure also wastes energy in compressing an otherwise unnecessarily large quantity of steam. Another effect of early cutoff is that the valve is moving quite slowly at the cutoff point, and this causes 'wire drawing' of the steam, another wasteful thermodynamic effect visible on an indicator diagram.

These inefficiencies drove the widespread experimentation in poppet valve gears for locomotives. Intake and exhaust poppet valves could be moved and controlled independently of each other, allowing for better control of the cycle. In the end, not a great number of locomotives were fitted with poppet valves, but they were common in steam cars and lorries, for example virtually all Sentinel lorries, locomotives and railcars used poppet valves. A very late British design, the SR Leader class, used sleeve valves adapted from internal combustion engines, but this class was not a success. In stationary steam engines, traction engines and marine engine practice, the shortcomings of valves and valve gears were among the factors that lead to compound expansion. In stationary engines trip valves were also extensively used.

Valve Gear Designs

Valve gear was a fertile field of invention, with probably several hundred variations devised over the years. However, only a small number of these saw any widespread use. They can be divided into those that drove the standard reciprocating valves (whether piston valves or slide valves), those used with poppet valves, and stationary engine trip gears used with semi-rotary Corliss valves or drop valves.

Reciprocating Valve Gears

- Gab or hook gear - used on earliest locomotives. Allowed reversing but no control of cutoff.

Link gears

Two eccentrics joined by a curved or straight link. A simple arrangement which works well at low speed. At high speed, a Walschaerts-type gear is said to give better steam distribution and higher efficiency.

- Stephenson valve gear - most common valve gear in 19th century, normally inside the locomotive frame.
- William T. James valve gear 1832, used first by Baltimore and Ohio Railroad, USA.
- Allan valve gear, a straight-link valve gear. It is fitted to 0-4-0WT "Dolgoch", preserved on the Talyllyn Railway.
- Gooch valve gear.

Radial Gears

Both components of the motion come from a single crank or eccentric. A problem with this arrangement (when applied to locomotives) is that one of the components of the motion is affected by the rise and fall of the locomotive on its springs. This probably explains why radial gears were largely superseded by Walschaerts-type gears in railway practice but continued to be used in traction and marine engines.

- Hackworth valve gear invented by John Wesley Hackworth in 1859.
- Joy valve gear - a design used extensively on the L&YR and LNWR in England, and elsewhere. A preserved example is LNWR G2a Class number 49395.
- Marshall valve gear - a modified Hackworth gear, patented in 1879 by Marshall, Sons & Co..
- Southern valve gear - Briefly popular in the United States around 1920. It combined elements of the Walschaerts and Baker patterns.

Walschaerts-type Gears

One component of the motion comes from a crank or eccentric. The other component comes from a separate source, usually the crosshead.

- Walschaerts or Heusinger valve gear - most common valve gear on later locomotives, normally externally mounted.
- Baker valve gear - fairly common in the United States, it had no sliding parts.

- Deeley valve gear - fitted to several express locomotives on the Midland Railway. The combination levers were driven, as normal, from the crossheads. Each expansion link was driven from the crosshead on the opposite side of the engine.
- Young valve gear - used the piston rod motion on one side of the locomotive to drive the valve gear on the other side. Similar to the Deeley gear, but with detail differences.
- Baguley valve gear - used by W.G. Bagnall.
- Bagnall-Price valve gear - a variation of Walschaerts used by W.G. Bagnall. This gear is fitted to Bagnall 3023 and 3050, both preserved on the Welsh Highland Railway.
- J. T. Marshall seems to have designed at least two different modifications of Walschaerts gear.
  - One was relatively conventional.
  - The other was very complex and drove separate valves on top of the cylinder (for admission) and underneath the cylinder (for exhaust). After the inventor's death, this gear was fitted experimentally to Southern Railway N Class locomotive number 1850, the work taking from 16 October 1933 to 3 February 1934; but it failed on 22 March 1934. Since the inventor was unable to modify the design, the valve gear was replaced by standard Walschaerts gear between 24 March and 11 April 1934.
- Isaacson's patent valve gear - a modified Walschaerts gear. It was fitted to the Garstang and Knot-End Railway's 2-6-0T *Blackpool.*

Poppet Valve Gears

- Caprotti valve gear, British Caprotti valve gear
- Hugo Lentz, oscillating-cam valve gear, rotary-cam valve gear
- Franklin oscillating-cam valve gear
- Franklin rotary-cam valve gear
- Reidinger valve gear

Conjugating Gears

These enable a 3-cylinder or 4-cylinder locomotive to be built with only two sets of valve gear. The best known is Gresley conjugated valve gear, used on 3-cylinder locomotives. Walschaerts gear is usually used for the two outside cylinders. Two levers connected to the outside cylinder valve rods drive the valve for the inside cylinder. Harold

Holcroft devised a different method for conjugating valve gear by linking the middle cylinder to the combination lever assembly of an outside cylinder, creating the Holcroft valve gear derivative. On a 4-cylinder locomotive the arrangement is simpler. The valve gear may be inside or outside and only short rocking-shafts are needed to link the valves on the inside and outside cylinders.

Bulleid Chain-driven Valve Gear

he Bulleid chain-driven valve gear is a type of steam locomotive valve gear designed by Oliver Bulleid for use on his Pacific (4-6-2) designs during the Second World War. They were peculiar to the Southern Railway in Britain, and were based upon motor-vehicle practice in an attempt to create a compact, but efficient design with a minimum of service requirements.

Design Principles

Oliver Bulleid's decision to have three cylinders, all driving the middle coupled axle of his Merchant Navy, West Country and Battle of Britain Classes, gave rise to several problems. As each cylinder was to have its own separate valve gear, this left very little space for the conventional inside set of motion. This prompted Oliver Bulleid to design a new miniaturised Walschaerts motion layout that was compact enough to enclose the whole system in a casing. All three sets of valve gear were worked from an auxiliary three-throw crankshaft. A Morse inverted tooth drive chain ran horizontally from a triple sprocket on the driving axle to an idler gear, from which a second chain ran downwards to the crankshaft. By the use of two chains the valve gear was isolated from the movement of the driving axle on its springs. The idler gear could be moved to give both chains the correct tension, but in practice only the vertical chain would be adjusted.

The auxiliary crankshaft drove both eccentric rods and combination levers displacing the piston valves; these were of the outside admission type. The valve heads were connected by a double girder arrangement, each actuated by a vertical rocking shaft located between the two girders and midway between the two heads, with a further connecting link pivoted just behind the valve head nearest the front. A sealed oscillating shaft drove the offset upper rocker arm and link, and the whole assembly worked inside the exhaust space of the steam chest. The advantage of having admission steam working on the outside faces of the valves meant that the volume under maximum pressure was completely sealed at the ends with no glands susceptible to leakage.

Valve motion and the inside connecting rod were enclosed in an oil bath consisting of a vertical steel box located between the main frame members, containing 40 gallons (182 litres) of oil. About 2 inches (50 mm) depth of oil lay in the bath; the inside big end was splash-lubricated, and two pumps sprayed oil through perforated pipes over the various valve motion pins. None of this was particularly revolutionary, being borrowed from internal-combustion engine practice and for use with steam it was established practice for steam motors at the Sentinel Waggon Works. It was thought that the arrangement would obviate the daily need to oil all moving parts and as they were protected from the elements, they should be able to run 100,000 miles (160,000 km) without attention. It was this consideration that meant the continued use of the system, albeit in modified form, on Bulleid's Leader Class.

Problems

In practice, the arrangement had a number of disadvantages. Cracks developed in the oil bath casing due to incorrect welding procedure. Condensation caused corrosion, and oil leaked out through inadequate seals, causing wheelslip and fire hazards. Valve timing was highly unpredictable; this has been attributed to chain stretch, although Bulleid claimed that it was allowed for. The chains were replaced as part of a valve and piston examination at around 30,000 to 36,000 miles, by which time they were considerably worn. This wear, in the chain pins and sprockets, could result in an elongation of the chains by as much as 6 inches. Bulleid asserted that a slack of 3 inches would be absorbed by the chain coming under load, with the remaining valve irregularity corrected by adjusting the cut-off.

Whilst chain-stretch is widely considered to be the cause of the gear's inefficiencies, another likely cause may have been geometrical due to the sequential proportions of levers, especially the final rockers transmitting the limited movement of the "miniaturised" valve gear to the union link driving the long travel piston valve. This "anti-lever" alone would have made accurate valve timing difficult to achieve, giving rise to stress in the drive mechanism, the symptom of which would be an increase in the likelihood of chain-stretch. The maximum throw of the expansion links was 2.25 inches (5.7 cm). To provide the maximum valve travel of 6 inches (15.2 cm) the rocker arms multiplied the motion of the auxiliary crankshaft in the ratio 3:8. The effects of wear in the pin joints, also multiplied by the same ratio, caused valve openings to vary between cylinders even if the driver maintained the same cut-off.

A further complication was the unpredictable behaviour of the Eastleigh type of steam reverser employed. If this reverser misbehaved under the fluctuating pressures present in the steam chest, the locomotive seemed to have a will of its own. For example, if the reverser dropped into full gear, the slackness in the motion would make the valves overrun their designed full travel inside the cylinders. This would result in the locomotive taking off like an unleashed race horse, high coal consumption, throwing the fire out of the chimney and running the risk of a violent high-speed slipping.

These problems combined contributed to the gear's eventual replacement by three separate sets of Walschaerts gear mounted in the conventional manner.

Bagnall-Price Valve Gear

Bagnall-Price valve gear is a type of steam engine valve gear developed at locomotive manufacturer W.G. Bagnall as an alternative to the more common Walschaerts valve gear. The gear was patented in 1903 by W.G. Bagnall and T. S. Price, the manager of the works.

Layout

The valves are driven from an eccentric cam mounted on one of the driving axles and linkage from the crosshead. The driving axle cam drives a *rocking shaft* that is mounted inside a cylindrical sleeve (known as the *rocking shaft bush*) above the cylinder. A drop lever and connecting link from the crosshead oscilates the sleeve to control the lap and lead of the cylinder.

Comparison to Walschaerts

W.G. Bagnall claimed that Bagnall-Price valve gear eliminated some of the complexity of Walschaerts gear, specifically the combination lever and return gear; this in turn meant the externally mounted gear had greater ground clearance which was an advantage in industrial locations where lineside obstructions could damage the gear of passing locomotives.

Baguley Valve Gear

The Baguley valve gear is a type of steam engine valve gear invented by Mr E. E. Baguley the draftsman of the W.G. Bagnall company of locomotive manufacturers. It was patented in 1893.

The valve gear is used to operate the inlet and outlet valves of reciprocating steam locomotives. It uses a rocking sector whose motion is driven via a link from a crankpin between the coupling and connecting rods. The ends of this link are circular and offset from the pivots. On these circular bosses the sector is oscillated by a second link from the rod to an extension to the rear of the sector.

Baker Valve Gear

Baker valve gear was the main competitor to Walschaerts valve gear for steam locomotives in the United States. Strictly speaking it was not a valve gear but a variable expansion mechanism adapted to the Walschaerts layout replacing the expansion link and sliding die block.

The Baker arrangement used more pivot bearings or *pin joints*, but avoided the die slip inherent to the expansion link, with the aim of lessening wear and the need for service; it could also facilitate longer valve travel.

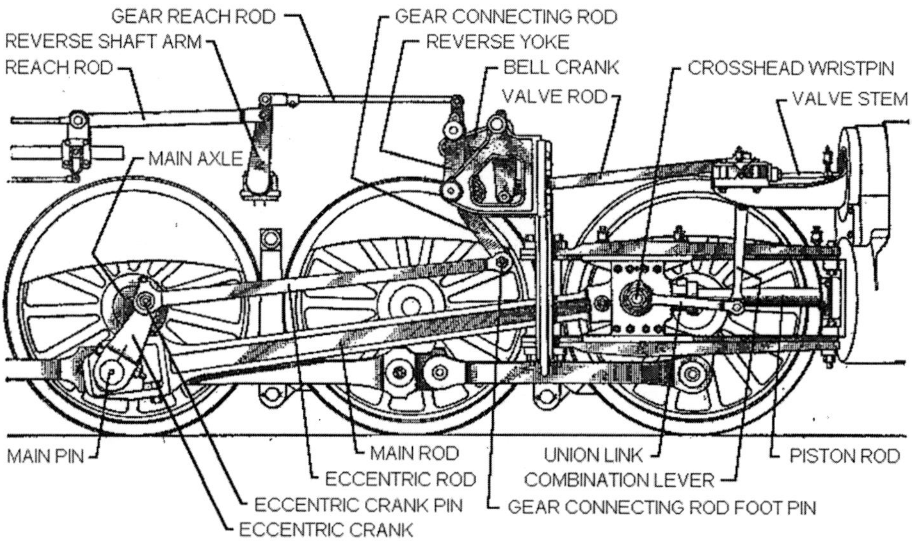

Figure: Baker Valve Gear Assembly

History

In the very early 1900s there were many efforts to create a new valve gear to replace the by-then-standard Walschaerts valve gear. In the United States the Young, Southern, Caprotti and Franklin patterns were used on a few classes, but only the Baker pattern won more than limited acceptance. The design originated in the A.D. Baker Company, of Swanton, Ohio; a builder of steam traction engines. The idea came from an employee called Gifford but was developed by the Baker company with the first patents being issued in 1903.

The Baker Locomotive Valve Gear was produced by the Pilliod Co. of Swanton, Ohio. Subsequent versions were produced up to the end of steam service. It was particularly popular on the Norfolk and Western Railway, and almost all later N&W engines used it (notably NW 611). Other extensive users included the Chesapeake and Ohio Railroad and the Nickel Plate Road. The New York Central Railroad and the Baltimore and Ohio Railroad also had large classes which used Baker gear. There was always heated debate about the claimed advantages of Baker gear, the main criticism being the number of pin joints and possible lost motion. Western United States and British railroads tended to prefer to continue with the Walschaerts pattern, as did the Pennsylvania Railroad. In Britain Baker gear was popular amongst model engineers but in full-size practice the length of the yoke and the width of the assembly may have been difficult to accommodate within the restricted loading gauge.

Operation

The Baker valve gear replaces the expansion link of the Walschaerts gear with an assembly of levers and links which produces the same effect of allowing continuous variation valve travel. The remainder of the gear is the same, so that the return crank and combination lever take the same form, although the proportions are usually modified. The Pilliod Co. sold the gear as a single assembly which was mounted on the frame in the location ordinarily occupied by the Walschaerts expansion link.

The Baker mechanism consisted of the following parts:

- A frame which replaced the mounting for the expansion link. It came in two basic forms, depending upon whether it was attached to the locomotive frame at both ends or only at the front. In either case it had two pivots: one forward and at the top of the frame, and one below and to the rear.
- A U-shaped reversing yoke which tilted back and forth to control reversing and cutoff. The ends of the yoke's arms pivoted on the rear frame mount, and there was also another pivot mounted near the base of the arm. The yoke was connected at the top to the reach rod leading to the reverser.
- A pair of swing links which hung from the upper pivot on the yoke.
- A main link which pivoted in the middle from the lower end of the swing links and whose lower end was connected to the eccentric rod (which in turn connects to the return crank on

the driver, as in the Walschaert gear). This link had a characteristic "J"-shape.

- A bellcrank whose centre pivot was the upper, forward mount on the frame, and whose ends were attached to the top of the main link and the radius rod (the connection to the combination arm and thence to the valve).

The parts were arranged so that when the yoke was centred, the connection of the main link to the bellcrank was in a line with the two pivots on the yoke. At this point, the back and forth motion of the lower end of the main link left the top relatively motionless as main link swung back and forth on the swing links.

In forward motion, the yoke was pushed forward, so that its upper pivot was in front of the bell crank-main link connection. Moving the eccentric arm back and forth lowered and raised the top of the main link. This motion was translated by the bellcrank into back and forth motion of the radius rod. The angle of the yoke controlled the relative motion, and therefore the cutoff; tilting the yoke backwards reversed the motion of the radius rod.

Bristol Centaurus

The Centaurus was the final development of the Bristol Engine Company's series of sleeve valve radial aircraft engines, an 18-cylinder, two-row design that eventually delivered over 3,000 hp (2,200 kW). It was one of the largest piston aircraft engines to enter production, and was introduced into service use towards the end of World War II. The Royal Navy Historic Flight currently operates a Hawker Sea Fury, powered by a Bristol Centaurus.

Design and Development

Like most Bristol Engines designs, the Centaurus was based on the mechanicals of an earlier design, in this case the "classic" 5.75 in (146 mm) piston from their original 1918 Jupiter. The Jupiter piston was still in use in the contemporary 14-cylinder Hercules, which was being brought into production during the design of the Centaurus. The Centaurus had a cylinder capacity of 3,270 in^3 (53.6 l), making it one of the largest piston aircraft engines to enter production, while that of the Hercules was 2,364 cubic inches (38.7 l). The nearly 40% higher capacity was achieved by increasing the stroke from 6.5 to 7 in (170 to 180 mm), and by changing to two rows of nine cylinders instead of two rows of seven, but the overall diametre of the Centaurus was only just over 6% greater.

While Bristol maintained the Centaurus dated from 1938 (the year it was originally type-tested), production could not start until 1942 owing to the need to get the Hercules into production and improve the reliability of the entire engine line. Nor was there any real need for the larger engine at this early point in the war, when most military aircraft designs were intended to mount engines of 1,000 hp or a little more. The Hercules' approximately 1,500 hp was simply better suited to the existing airframes then in production.

In fact, the Centaurus did not see any use until near the end of the war, first appearing on an upscaled version of the Vickers Wellington, the Warwick. Other wartime, or just-postwar, applications included the Bristol Brigand and Buckmaster, Hawker Tempest and Sea Fury, and the Blackburn Firebrand and Beverley. The engine also saw post-war use in civilian airliners, including the ill-fated Bristol Brabazon. By the end of the war in Europe, around 2,500 examples of the Centaurus had been produced by Bristol. A projected enlarged capacity version of the Centaurus was designed by Roy Fedden, cylinders were produced for this engine but it was never built. Known as the Bristol Orion (a name used previously for a variant of the Jupiter engine) this development was also a two-row, 18 cylinder sleeve valve engine with the displacement increased to 4,142 cubic inches (67.9 l).

Variants

The Centaurus was produced in 34 distinct variants ranging from the 2,000 hp (1,490 kW) Centaurus I to the 2,405 hp (1,793 kW) Centaurus 663 for the Airspeed Ambassador airliner. The most powerful variants were the 2,625 hp (1,957 kW) Centaurus 170, 173, 660, 661 and 662.

Blowoff Valve

A blowoff valve (BOV) or dump valve is a pressure release system present in most turbocharged engines. Its purpose is to prevent compressor surge, and reduce wear on the turbocharger and engine. Blowoff valves relieve the damaging effects of compressor "surge loading" by allowing the compressed air to vent to the atmosphere, making a distinct hissing sound, or recirculated into the intake upstream of the compressor inlet.

Definitions

A compressor bypass valve (CBV), also known as a pressure relief valve or diverter valve, is a manifold vacuum-actuated valve designed to release pressure in the intake system of a turbocharged vehicle when the throttle is lifted or closed. This air pressure is re-circulated back into the non-pressurized end of the intake (before the turbo) but after the mass airflow sensor.

A blowoff valve, (sometimes "hooter valve", not to be confused with a bypass valve) performs the same task but releases the air into the atmosphere instead of recirculating it. The blowoff action produces a range of distinctive hissing sounds, depending on the exit design. Some blowoff valves are sold with a trumpet-shaped exit that intentionally amplifies the sound. Some turbocharged vehicle owners may purchase a blowoff valve solely for the auditory effect even when the function is not required by normal engine operation. Motor sports governed by the FIA have made it illegal to vent unmuffled blowoff valves to the atmosphere.

Blowoff valves are used to prevent compressor surge, a phenomenon that readily occurs when lifting off the throttle of an unvented, turbocharged engine. When the throttle plate on a turbocharged engine closes, the high pressure air in the intake system is trapped by the throttle and a pressure wave is forced back into the compressor. The compressor wheel slows rapidly and may even stall, and the driver will notice a fluttering air sound. The rapid slowing or stalling stresses the turbo and imparts severe turbo lag if the driver accelerates immediately after the surge event.

Disadvantages

In the case where a mass airflow sensor (MAF) is used and is located upstream from the blowoff valve, the engine control unit (ECU) will metre out excess fuel because the atmospherically vented air is not subtracted from the intake charge measurements. The engine then briefly operates with a fuel-rich mixture after each valve actuation.

The rich mixing can lead to hesitation or even stalling of the engine when the throttle is closed, a situation that worsens with higher boost pressures. Occasional events of this type may be only a nuisance, but frequent events can eventually foul the spark plugs and destroy the catalytic converter, as the inefficiently combusted fuel produces soot (excess carbon) and unburned fuel in the exhaust flow can produce soot in the converter and drive the converter beyond its normal operating temperature range.

An alternative method for utilising both a MAF and a blowoff valve is to have the MAF located down stream between the intercooler and the throttle plate. This is known as Blow-through rather than the traditional Draw-through set up. Care must be taken as to the position of the MAF to prevent damage to the sensitive element. For example, on a SR20DE motor, the MAF must be at least 12" from the throttle plate, and the blowoff valve must be 6" from the MAF sensor. By using a draw-through method, the MAF won't be affected by the blowoff valve opening as the pressure is vented before the air reaches the MAF.

One approach used to mitigate the problem has been to reduce the boost pressure, which reduces the required venting volume and yields less charge over-calculation by the ECU. The air can also be recirculated back into the intake, a typical stock setup for cars with an upstream MAF sensor. The situation can also be corrected by switching the fuel metring system over to a manifold absolute pressure sensor, a conversion that usually requires a compatible aftermarket ECU or piggy-back fuel controller. The MAP sensor monitors the absolute pressure in the manifold at all times and will correctly detect the change that occurs when the valve vents, allowing the ECU to reduce fuel metring accordingly.

Operation

A blowoff valve is connected by a vacuum hose to the intake manifold *after* the throttle plate. When the throttle is closed, the relative manifold pressure drops to below zero bar and the resulting pressure differential operates the blowoff valve's piston. The excess pressure from the turbocharger is then vented into the atmosphere or recirculated into the intake upstream of the compressor inlet.

Multi-valve

In automotive engineering a multi-valve or multivalve engine is one where each cylinder has more than two valves. A multi-valve engine has better breathing and can operate at higher revolutions per minute (RPM) than a two-valve engine, delivering more power.

Multi-valve Rationale

Multi-valve Design: A multi-valve design typically has three, four, or five valves per cylinder to achieve improved performance. Any four-stroke internal combustion engine needs at least two valves per cylinder: one for *intake* of air and fuel, and another for *exhaust* of combustion gases. Adding more valves increases valve area and improves the flow of intake and exhaust gases, thereby enhancing combustion, volumetric efficiency, and power output. Multi-valve geometry allows the spark plug to be ideally located within the combustion chamber for optimal flame propagation. Multi-valve engines tend to have smaller valves that have lower reciprocating mass, which can reduce wear on each cam lobe, and allow more power from higher RPM without the danger of valve bounce.

Some engines are designed to open each intake valve at a slightly different time, which increases turbulence, improving the mixing of air and fuel at low engine speeds. More valves also provide additional cooling to the cylinder head.

The disadvantages of multi-valve engines are an increase in manufacturing cost and a potential increase in oil consumption due to the greater number of valve stem seals. Some SOHC multi-valve engines (such as the Mazda B8-ME) use a single fork-shaped rocker arm to drive two valves (generally the exhaust valves) so that fewer cam lobes will be needed in order to reduce manufacturing costs.

- *Three-valve cylinder head*: This has a single large exhaust valve and two smaller intake valves. A three-valve layout allows better breathing than a two-valve head, but the large exhaust valve results in a lower RPM limit. The manufacturing cost for this design can be lower than for a four-valve design. This design was common in the late 1980's and early 1990's, but has been largely superseded by four and five-valve designs.
- *Four-valve cylinder head*: This is the most common type of multi-valve head, with two exhaust valves and two similar (or slightly larger) inlet valves. This design allows similar breathing as compared to a three-valve head, and as the small exhaust valves allow high RPM, this design is very suitable for high power outputs.
- *Five-valve cylinder head*: Less common is the five-valve head, with two exhaust valves and three inlet valves. All five valves are similar in size. This design allows excellent breathing, and, and as every valve is small, high RPM and very high power outputs are available. It has been questioned whether a five-valve configuration gives a worthwhile benefit over four-valve designs. In theory, a five-valve design should have a higher maximum RPM, and the three inlet ports should give both good cylinder-filling and high turbulence (both desirable traits). After making the five-valve Genesis engine for several years, Yamaha returned to the more cost-effective four-valve design.

Alternative Technologies

Turbocharging and supercharging are technologies that also improve engine breathing, and can be used instead of, or in conjunction with, multi-valve engines. Same applies to variable valve timing and variable intake manifolds. Rotary valves also offer improved engine breathing and high rev performance but these were never very successful. Cylinder head porting, as part of engine tuning, is also used to improve engine performance.

Cars and Trucks

Before 1914

The first motorcar in the world to have an engine with two overhead camshafts and four valves per cylinder was the 1912 Peugeot L76 Grand Prix race car. Its 7.6-litre monobloc straight-4 with modern hemispherical combustion chambers produced 148 bhp (0.32 bhp per cubic inch). In April 1913, on the Brooklands racetrack in England, a specially built L76 called *"la Torpille"* (torpedo) beat the world speed record of 170 km/h. Robert Peugeot also commissioned the young Ettore Bugatti to developed a GP racing car for the 1912 Grand Prix. This chain-driven Bugatti Type 18 had a 5-litre straight-4 with SOHC and three valves per cylinder (two inlet, one exhaust). It produced appr. 100 bhp at 2800 rpm (0.30 bhp per cubic inch) and could reach 99 mph. The three-valve head would later be used for some of Bugatti's most famous cars, including the 1922 Type 29 Grand Prix racer and the legendary Type 35 of 1924. Both Type 29 and Type 35 had a 100 bhp 2-litre SOHC 24-valve NA straight-8 that produced 0.82 bhp per cubic inch.

Between 1914 and 1945

In 1916 US automotive magazine *Automobile Topics* described a four-cylinder, four-valve-per-cylinder car engine made by Linthwaite-Hussey Motor Co. of Los Angeles, CA, USA: *"Firm offers two models of high-speed motor with twin intakes and exhausts."*.

Early multi-valve engines in T-head configuration were the 1917 *Stutz* straight-4 and 1919 *Pierce-Arrow* straight-6 engines. The standard flathead engines of that day were not very efficient and designers tried to improve engine performance by using multiple valves. The Stutz Motor Company used a modified T-head with 16 valves, twin-spark ignition and aluminium pistons to produce 80 bhp (59 kW) at 2400 rpm from a 360.8 cid (5.8-litre) straight-4 (0.22 bhp per cubic inch). Over 2300 of these powerful early multi-valve engines were built. Stutz not only used them in their famous Bearcat sportscar but in their standard touring cars as well. In 1919 Pierce-Arrow introduced its 524.8 cid (8.6-litre) straight-6 with 24 valves. The engine produced 48.6 bhp (0.09 bhp per cubic inch) and ran very quietly, which was an asset to the bootleggers of that era.

Multi-valve engines continued to be popular in racing and sports engines. Robert M. Roof, the chief engineer for Laurel Motors, designed his multi-valve *Roof Racing Overheads* early in the 20th century. Type A 16-valve heads were successful in the teens, Type B was offered in 1918 and Type C 16-valve in 1923. Frank Lockhart drove a Type C overhead cam car to victory in Indiana in 1926.

Bugatti also had developed a 1.5-litre OHV straight-4 with four valves per cylinder as far back as 1914 but did not use this engine until after World War I. It produced appr. 30 bhp (22.4 kW) at 2700 rpm (15.4 kW/litre or 0.34 bhp/cid). In the 1920 Voiturettes Grand Prix at Le Mans driver Ernest Friderich finished first in a Bugatti Type 13 with the 16-valve engine, averaging 91.96 km/h. Even more successful was Bugattis clean sweep of the first four places at Brescia in 1921. In honour of this memorable victory all 16-valve-engined Bugattis were dubbed *Brescia*. From 1920 through 1926 about 2000 were built.

A.L.F.A. 40/60 GP was a fully working early racing car prototype made by the company now called Alfa Romeo. Only one example was built in 1914, which was later modified in 1921. This design of Giuseppe Merosi was the first Alfa Romeo DOHC engine. It had four valves per cylinder, 90-degree valve angle and twin-spark ignition. The GP engine had a displacement of 4.5-litre (4490 cc) and produced 88 bhp (66 kW) at 2950 rpm (14.7 kW/litre), and after modifications in 1921 102 bhp (76 kW) at 3000 rpm. The top speed of this car was 88-93 mph (140–149 km/h). It wasn't until the 1920s when these DOHC engines came to Alfa road cars like the Alfa Romeo 6C.

Peugeot had a triple overhead cam 5-valve Grand Prix car in 1921.

Bentley used multi-valve engines from the beginning. The Bentley 3 Litre, introduced in 1921, used a monobloc straight-4 with aluminium pistons, pent-roof combustion chambers, twin spark ignition, SOHC, and four valves per cylinder. It produced appr. 70 bhp (0.38 bhp per cubic inch). The 1927 Bentley 4½ Litre was of similar engine design. The NA racing model offered 130 bhp (0.48 bhp per cubic inch) and the 1929 supercharged 4½ Litre (Blower Bentley) reached 240 bhp

(0.89 bhp per cubic inch). The 1926 Bentley 6½ Litre added two cylinders to the monobloc straight-4. This multi-valve straight-6 offered 180-200 bhp (0.45-0.50 bhp per cubic inch). The 1930 Bentley 8 Litre multi-valve straight-6 produced appr. 220 bhp (0.45 bhp per cubic inch).

In 1931 the Stutz Motor Company introduced a 322 cid (5.3-litre) dual camshaft 32-valve straight-8 with 156 bhp (116 kW) at 3900 rpm, called DV-32. The engine offered 0.48 bhp per cubic inch. About 100 of these multi-valve engines were built. Stutz also used them in their top-of-the-line sportscar, the DV-32 Super Bearcat that could reach 100 mph (160 km/h).

The 1935 Duesenberg SJ Mormon Meteor's engine was a 419.6 cid (6.9-litre) straight-8 with DOHC, 4 valves per cylinder and a supercharger. It achieved 400 bhp (298.3 kW) at 5,000 rpm and 0.95 bhp per cubic inch.

The 1937 Mercedes-Benz W125 racing car used a supercharged 5.7-litre straight-8 with DOHC and four valves per cylinder. The engine produced 592-646 bhp (441.5-475 kW) at 5800 rpm and achieved 1.71-1.87 bhp per cubic inch (77.8-85.1 kW/litre). The W125 top speed was appr. 200 mph (322 km/h).

After 1945

The 1967 Cosworth DFV F1 engine, a NA 3.0-litre V8 producing appr. 400 bhp (298 kW; 406 PS) at 9,000 rpm (101.9 kW/litre), featured four valves per cylinder. For many years it was the dominant engine in Formula One, and it was also used in other categories, including CART, Formula 3000 and Sportscar racing.

Debuting at the 1968 Japanese Grand Prix in the original 300 PS (221 kW; 296 hp) 3.0-litre version the Toyota 7 engine participated in endurance races as a 5.0-litre (4,968 cc) non-turbo V8 with DOHC and 32-valves. It produced 600 PS (441 kW; 592 hp) at 8,000 rpm (88.8 kW/litre) and 55.0 kg·m (539 N·m; 398 ft·lbf) at 6,400 rpm.

The first mass-produced car using four valves per cylinder was the British Triumph Dolomite Sprint of 1973 which used a chain-driven SOHC 16-valve straight-4 producing 127 bhp (95 kW; 129 PS).

The 1975 Chevrolet Cosworth Vega featured a DOHC multi-valve head designed by Cosworth Engineering in the UK. This 122-cubic-inch straight-4 produced 110 bhp (82 kW; 112 PS) at 5600 rpm (0.90 bhp/cid; 41.0 kW/litre) and 107 lb·ft (145 N·m) at 4800 rpm.

The 1976 Fiat 131 Abarth (51.6 kW/litre), 1976 Lotus Esprit with

Lotus 907 engine (54.6 kW/litre, 1.20 bhp/cid), and 1978 BMW M1 with BMW M88 engine (58.7 kW/litre, 1.29 bhp/cid) all used four valves per cylinder. The BMW M88/3 engine was used in the 1983 BMW M635CSi and in the 1985 BMW M5.

The 1978 Porsche 935/78 racer used a twin turbo 3.2-litre flat-6 (845 bhp/630 kW@8,200 rpm; 784 Nm/578 ft.lbs@6,600 rpm). The water-cooled engine featured four valves per cylinder and output a massive 196.2 kW/litre. Porsche had to abandon its traditional aircooling because the multi-valve DOHC hampered aircooling of the spark plugs. Only two cars were built.

Ferrari developed their *Quattrovalvole* (or QV) engines in the 80s. Four valves per cylinder were added for the 1982 308 and Mondial *Quattrovalvole*, bringing power back up to the pre-FI high of 245 hp (183 kW). A very unusual Dino Quattrovalvole was used in the 1986 Lancia Thema 8.32. It was based on the 308 QV's engine, but used a split-plane crankshaft rather than the Ferrari-type flat-plane. The engine was constructed by Ducati rather than Ferrari, and was produced from 1986 through 1991. The Quattrovalvole was also used by Lancia for their attempt at the World Sportscar Championship with the LC2. The engine was twin-turbocharged and destroked to 2.65 litres, but produced 720 hp (537 kW) in qualifying trim. The engine was later increased to 3.0 litres and increased power output to 828 hp (617 kW). The 1984 Ferrari Testarossa had a 4.9-litre flat-12 with four valves per cylinder. Almost 7,200 Testarossa were produced between 1984 and 1991.

In 1985 Lamborghini released a Countach *Quattrovalvole*, producing 455 PS (335 kW; 449 hp) from a 5.2-litre (5167 cc) Lamborghini V12 engine (64.8 kW/litre).

The Mercedes-Benz 190E 2.3-16 with 16-valve engine debuted at the Frankfurt Auto Show in September 1983 after it set a world record at Nardo, Italy, recording a combined average speed of 154.06 mph (247.94 km/h) over the 50,000 km (31,000 mi) endurance test. The engine was based on the 2.3-litre 8-valve 136 hp (101 kW) unit already fitted to the 190- and E-Class series. Cosworth developed the DOHC light alloy cast cylinder head with four large valves per cylinder. In roadgoing trim, the 190 E 2.3-16 produced 49 hp (36 kW) and 41 ft•lbf (55 N•m) of torque more than the basic single overhead cam 2.3 straight-4 engine on which it was based offering 185 hp (138 kW) at 6,200 rpm (59.2 kW/litre) and 174 lb·ft (236 N·m) at 4,500 rpm. In 1988 an enlarged 2.5-litre engine replaced the 2.3-litre. It offered double valve timing chains to fix the easily snapping single chains on early 2.3 engines, and increased peak output by 17 bhp (12.5 kW) with a slight increase in torque. For homologation Evolution I (1989) and Evolution II (1990) models were produced that had a redesigned engine to allow for a higher rev limit and improved top-end power capabilities. The Evo II engine offered 235 PS (173 kW; 232 hp) from 2463 cc (70.2 kW/litre).

Saab introduced a 16-valve head to their 2.0-litre (1985 cc) straight-4 in 1984 and offered the engine with and without turbocharger (65.5 kW/litre and 47.9 kW/litre respectively) in the Saab 900 and Saab 9000.

The 1.6-litre (1,587 cc) 4A-GE Toyota engine was one of the earliest straight-4 engines to have both a DOHC 16-valve configuration (four valves per cylinder, two intake, two exhaust) and electronic fuel injection (EFI). The cylinder head was developed by Yamaha Motor Corporation and was built at Toyota's Shimayama plant. While originally conceived of as a two-valve design, in 1984 Toyota and Yamaha changed the 4A-GE to a four-valve after a year of evaluation. It produced 115-140 bhp/86-104 kW@6,600 rpm (54.2-65.5 kW/litre) and 148 Nm/109 lbft@5,800 rpm. To compensate for the reduced air speed of a multi-valve engine at low rpm, the first-generation engines included the T-VIS feature.

In 1986 Volkswagen introduced a multi-valved Golf GTI 16V. The 16-valve 1.8-litre straight-4 produced 139 PS (102 kW; 137 bhp) or 56.7 kW/litre, almost 25% up from the 45.6 kW/litre for the previous 8-valve Golf GTI engine.

The GM Quad 4 multi-valve engine family debuted early 1987. The Quad 4 was the first mainstream multi-valve engine to be produced by GM after the Chevrolet Cosworth Vega. The NA Quad 4 achieved 1.08 bhp (1 kW; 1 PS) per cubic inch (49.1 kW/litre). Such engines soon became common as Japanese manufacturers adopted the multi-valve concept.

Three Valves

The 1984 Honda CR-X and 1985 Honda Civic introduced Honda's 1.5-litre SOHC 12-valve straight-4 engines. Nissan's 1988–1992 SOHC KA24E engine had three valves per cylinder (two intakes, one exhaust) as well. Nissan upgraded to the DOHC after 1992 for some of the sports cars, including the 240SX.

Mercedes and Ford produce three-valve V6 and V8 engines, Ford claiming an 80% improvement in high RPM breathing without the added cost of a DOHC valve train. The Ford design uses one spark plug per cylinder located in the centre, but the Mercedes design uses two spark plugs per cylinder located on opposite sides, leaving the centre free to add a direct-to-cylinder fuel injector at a later date.

The 1989 Citroën XM was the first 3-valve diesel-engined car.

Four valves

Examples of SOHC four-valve engines include: the Honda F-series engines, D-series engines, all J-series engines, the R-series engines, the Mazda B8-ME, the Chrysler 3.5 L V6 engine.

The V12 engines of many World War II fighter aircraft also used a SOHC configuration with four valves for each cylinder.

The 1993 Mercedes-Benz C-Class (OM604 engine) was the first 4-valve diesel-engined car.

Five Valves

Peugeot had a triple overhead cam five-valve Grand Prix car in 1921.

In April 1988 an Audi 200 Turbo Quattro powered by an experimental 2.2-litre turbocharged 25-valve straight-5 rated at 478 kW/650 PS@6,200 rpm (217.3 kW/litre) set two world speed records at Nardo, Italy: 326.403 km/h (202.8 mph) for 1,000 km (625 miles) and 324.509 km/h (201.6 mph) for 500 miles.

Mitsubishi were the first to market a car engine with five valves per cylinder, with the 548 cc 3G81 engine in their Minica Dangan ZZ kei car in 1989.

Yamaha designed the five-valve cylinder head for the Toyota 4A-GE 20V 1991 Silvertop and 1995 Blacktop engine used in the MR2 and some Corollas. Yamaha also developed five-valve Formula One engines, the 1989 OX88 V8, 1991 OX99 V12, 1993 OX10 V10 and 1996 OX11 V10, but none of these were very successful.

Bugatti (EB110), Ferrari (F355 and F50), Volkswagen - Audi (Audi Quattro) - Skoda (Octavia vRS) and Toyota have all produced five-valve-engined vehicles.

Six Valves

In 1985 Maserati made an experimental 2.0-litre turbo V6 engine with six valves per cylinder (three intakes, three exhaust). It achieved 261 bhp (195 kW; 265 PS) at 7,200 rpm (97.5 kW/litre).

Pushrod

Although most multi-valve engines have overhead camshafts,

either SOHC or DOHC, a multivalve engine may be a pushrod overhead valve engine (OHV) design. Chevrolet has revealed a three-valve version of its Generation IV V8 which uses pushrods to actuate forked rockers, and Cummins makes a four-valve OHV straight six diesel, the Cummins B Series (now known as ISB).

Turbocharged

The 1981 Maserati Biturbo (with AM452 18-valve 2-litre V6 engine) was the first production car to use a multi-valve turbocharged engine.

Motorcycles

Examples of motorcycles with multivalve-engines include:

- 1914 Peugeot Grand Prix racer, 500 cc DOHC 8-valve parallel twin (top speed over 122 km/h).
- 1915 Indian board track racer, 61-cid (1.0-litre) OHV 8-valve V-twin.
- 1921 Triumph Ricardo 499 cc OHV 4-valve single-cylinder machine, copied by Rudge-Whitworth with their 1924 Rudge Four 350 cc OHV 4-valve single-cylinder machine, and 1929 Rudge Ulster 500 cc OHV 4-valve single-cylinder machine.
- 1972 Honda XL250 "pent-roof" SOHC 4-valve single-cylinder machine (the first mass-produced 4-valve motorcycle).
- 1977 Honda CB400 SOHC 6-valve parallel-twin.
- 1978 Honda CX500, a 498 cc OHV 8-valve V-twin, and 1978 Honda CBX1000, a 1,047 cc DOHC 24-valve straight-6 (105 bhp (78 kW; 106 PS)).
- 1979 -1992: Honda NR series, racing & production motorcycles with 8-valve-per-cylinder "oval-piston" V4 engines (actually 32-valve V8s with adjoining cylinders merged).
- 1985 Yamaha FZ750 motorcycle with DOHC 20-valve straight-4 Yamaha "Genesis" engine.
- 1998–2006 Yamaha YZF-R1 superbike with redesigned (more compact) "Genesis" engine. 2006 model delivered 180 bhp (134 kW; 182 PS) at 12,500 rpm (130.3 kW/litre).

The Yamaha XT660 single once had five valves per cylinder, but a subsequent redesign reduced the valve-count to four. The Aprilia Pegaso 650 single also started out with five valves, but current models only have four. The jointly developed BMW F650 single always had four valves.

Aircraft

Ettore Bugatti designed several multi-valve aircraft engines. The 1916 Bugatti U-16 1484.3 cid (24.32 L) SOHC 16-cylinder, consisting of two parallel 8-cylinder banks, offered 410 bhp (305 kW) at 2,000 rpm (12.5 kW/litre or 0.28 bhp/cid). Each cylinder had two vertical inlet valves and a single vertical exhaust valve, all driven by rocking levers from the camshaft. Other advanced World War I aircraft engines, such as the 1916 Maybach Mb.IVa that produced 300 bhp (224 kW; 304 PS) at altitude and the 1916 Benz Bz.IV with aluminium pistons and the 1918 Napier Lion (a 450 bhp 24-litre DOHC 12-cylinder), used two intake valves and two exhaust valves.

The V12 engines of many World War II fighter aircraft used a SOHC configuration with four valves for each cylinder.

An example of a modern multi-valve piston-engine for small aircraft is the Austro Engine AE300. This liquid-cooled turbocharged 2.0-litre (1,991 cc) DOHC 16-valve straight-4 diesel engine uses common rail direct fuel injection and delivers 168 bhp (125 kW; 170 PS) at 3,880 rpm (62.0 kW/litre). The propeller is driven by an integrated gearbox (ratio 1.69:1) with torsional vibration damper. Total power unit weight is 185 kg (408 lb).

Boats

In 1905 car builder Delahaye had experimented with a DOHC marine racing engine with six valves per cylinder. This Delahaye 'Titan' engine was a massive 5190 cid (85.0-litre) four-cylinder that produced 350 bhp (0.07 bhp/cid). It allowed the motor boat Le Dubonnet piloted by Emile Dubonnet to set a new world's speed record on water, reaching 33.80 mph (54.40 km/h) on the lake at Juvisy, near Paris, France.

An example of modern multi-valve engines for small boats is the Volvo Penta IPS Series. These joystick-operated seawater-cooled inboard diesel engines use combined charging (turbo and supercharger, except IPS450) with aftercooler, common rail fuel injection and DOHCs with hydraulic 4-valve technology. Propshaft power ranges from 248 to 850 bhp (185 to 634 kW; 251 to 862 PS) (highest efficiency 59.7 kW/litre for IPS400 3.7-litre straight-4 diesel). Multiple units can be combined.

Chapter 8

Pneumatic Valve Springs

Pneumatic valve springs are metal bellows filled with compressed air used as an alternative to the metal wire springs used to close valves in high-speed internal combustion engines. This system was introduced in the mid-1980s in Renault turbocharged 1.5 litre Formula One engines.

Concept

Racing engines often fail at high rotational speeds because mechanical springs are unable to retract the valves quickly enough to provide clearance for the piston. Renault's pneumatic valve technology replaced steel springs with light weight compressed air bellows. These could retract valves more quickly and reduce the possibility of piston-valve interference as long as pressure could be maintained. Additionally, the amount of seat tension required to keep a coil sprung valve under control results in greater peak lift loading, with added stress to the entire valvetrain as a result. Pneumatic systems, sharing a common reservoir of pressure retain a more static level of force, controlling the valve effectively, without any attendant peak lift load increase.

Bellows is a misleading word, since this would suggest some sort of concertina shaped cylinder. This is not the case, and a bellows of this design would fail due to metal fatigue at the roots and crests. Instead, it is simply a piston and cylinder, similar to a small pneumatic ram. Where the tappet bore where a hydraulic tappet would normally reside, becomes the cylinder, and the retainer assembly becomes the piston. Pressurised air(nitrogen) is pumped into this cylinder which then causes the piston/ retainer to rise to the top of cylinder, causing the valve to form an airtight seal with the seat. The compressed gas then becomes the spring so to speak, and does not have the same traits as springs do at elevated rpm. A small light spring is sometimes fitted between the piston, and retainer so that when the system is switched off, the spring forces the piston down against the bottom of bore, and forces the retainer upwards. This ensures that no crown to valve contact occurs when shut down.

Pneumatic Valve Technology in Racing

Pneumatic valve springs gave Renault an advantage with its turbocharged engines, often said to be the most powerful. However, reliability and poor handling of their chassis kept the cars from success until 1989 when Renault provided Williams with its V10 engine to begin a winning streak. Renault won the constructors' championship with Williams in 1992, 1993, 1994, 1996 and 1997, and with Benetton in 1995. Nigel Mansell, Alain Prost, Michael Schumacher, Damon Hill and Jacques Villeneuve each won a Driver's championship in the 1990s with Renault engines. Pneumatic valve springs are also found in several Moto GP motorcycle engines, debuting in 2002 with the Aprilia RS3 Cube. In 2005, Team Roberts was the first to use pneumatic valves full time in their uncompetitive KTM powered bike. Today, almost all of the MotoGP teams use pneumatic valve technology on their bikes, including Yamaha, Suzuki and Honda. Ducati uses a desmodromic design.

Future Valve Technology

While pneumatic valve springs have become standard in Formula One engines, Renault has been researching computer controlled electromagnetic suspension valves using no camshaft, to reduce moving parts while improving valve control.

Vacuum Switch

Prior to effective engine control unit computers, engine vacuum was used for many functions in an automobile. Vacuum switches were employed to regulate this flow, and were commonly controlled by temperature, solenoids, mechanically, or directly. They operated vacuum motors, other vacuum switches and other devices. The internal combustion engine in a common automobile produces almost 20 inches (51 cm) of vacuum, and this pressure differential may be harnessed for many uses. Engine vacuum is also the best direct source of information on the engine's load. Most delay valves have a one-way function, where there is either no restriction or no movement in one direction.

Types

There are several common types of vacuum switches. Many are completely useless outside of their exact designed role.

Check Valve

A valve that only allows the vacuum signal to move in one direction. Often used with vacuum reservoirs.

Delay Valve

A vacuum delay valve is a valve with a small orifice, which delays a vacuum signal. These are commonly used in automobiles to alter the behaviour of a vacuum signal. Delay valves are usually colour-coded to their function.

Coolant Temperature Override

Also referred to as a Temperature Vacuum Switch (TVS) or Thermal Vacuum Valve (TVV). The CTO switch measures coolant temperature, usually from a passage in the intake manifold. It was commonly used starting in the 1960s for switching the vacuum advance from ported vacuum to manifold vacuum at high engine temperatures to promote cooler idling. As emissions controls expanded, it gained many other uses. They commonly have between two and five vacuum ports.

Thermostatically-controlled Air Cleaner System

The factory air cleaner on carbureted vehicles has a switch controlled by a bi-metallic strip that ultimately selects cool or preheated air. This valve is usually open on one side to the air space within the air filter, and is both temperature controlled, and bleed style. Its function is to keep the air going in to the carburetor above a certain temperature to prevent icing.

Non-linear Vacuum Regulator

A NLVR regulates vacuum signal so it is between two vacuum source levels. As one signal increases the other regulator valve switches to another vacuum source.

Bleed Valve

A bleed valve controls a metred orifice that adjusts to allow air into the line to weaken a vacuum signal. They are commonly controlled by coolant temperature or under hood air temperature, but may be part of the HVAC system.

Vacuum Booster

A vacuum booster uses one vacuum signal, usually ported, to control another signal, usually manifold, that is stronger. It may also use a venturi effect.

Positive Crankcase Ventilation Valve

The positive crankcase ventilation (PCV) is a one-way valve that ensures continual refreshment of the air inside the engine's crankcase.

Vacuum Electrical Switch

A vacuum switch that turns an electrical connection on or off. Used for early electronic control modules to determine engine load or function.

Vacuum Solenoid

An electrical solenoid that switches a vacuum signal on or off.

Manual Vacuum Switch

Operator controlled switches were often used to control vacuum powered devices before the advent of robust automotive electrical systems or reliable mechanical linkages. They consisted of a variety of shapes, sizes, and configurations much like electrical switches.

They are commonly found as window and HVAC controls.

Electrical Switches

The term "vacuum switch" can also refer to a specialised type of electrical switch in which the contacts are enclosed in a vacuum envelope. This allows the switching of very high voltages with a compact switch, while preventing arcing across the contacts.

Valve Float

Valve float is an adverse condition which occurs when the poppet valves on an internal combustion engine valvetrain do not remain in contact with the camshaft lobe during the valve closure phase of the cam lobe profile. This reduces engine efficiency and performance and potentially increases engine emissions.

Similar Conditions

Valve loft is utilising a controlled float to increase lift and duration of the valve open cycle. In some motorsports there are rules that limit camshaft lift; therefore, provoking this type of exploitation. By properly optimizing the system, one can actually achieve this without undue stresses added to the camshaft lobes and tappets (lifters).

Valve bounce is a related condition where the valve does not stay seated, due to the combined effects of the valve's inertia and resonance effects of metallic valve springs that effectively reduce the closing force, and allow the valve to re-open partially.

Remedies

Stiffer valve springs can help prevent valve float and valve bounce, but only at the expense of increased friction losses and higher stresses in the valvetrain system. Various techniques have been used to offset the effect of stiffer springs, such as dual-spring and progressive-sprung valves, roller-tipped tappets, and pneumatic valves.

Alternative Solutions

Italian motorcycle manufacturer Ducati uses the desmodromic (springless) valve system to counter this problem and allow for higher engine speeds. The system consists of a mechanical lifter mechanism that uses a second rocker arm to push the valve closed. Formula One racing engine manufacturers use a pneumatic system to close the valves to allow for very high RPM without valve float.

Valve Guide

In most types of reciprocating engines, a valve guide is provided for each poppet valve in the cylinder head. Along with the valve spring, it serves to positively locate the valve so that it may make proper contact with the valve seat. A valve guide is a tube-shaped piece of metal, pressed into the cylinder head, with the valve reciprocating inside it. Guides serve also to conduct heat from the combustion process out from the exhaust valve and into the cylinder head where it may be taken up by the cooling system. Bronze is commonly used, as is steel; a balance between stiffness and wear on the valve is essential to achieve a useful service life.

Aircraft Engines

In aircraft piston engines, the valve guides are typically shaped in a tube with a flare at one end, so as to resemble in section, the letter "T". Their replacement involves removing the worn part by driving it out with a hammer and specifically shaped punch. Installation involves heating the cylinder (with its integral head) and freezing the valve guide so as to ease insertion, and then driving the new guide in (quickly) with a hammer. Once the parts return to room temperature the new valve guide will be solidly in place and ready to be reamed and honed to proper diametre...

Oil Seals

As valve guides wear, their ability to positively locate the valve to the valve seat decreases. As the valves lose their ability to seal the combustion chamber properly, the engine can lose performance and start to burn oil, leaking from the top of the cylinder head into the intake and exhaust manifolds. Modern engines are fitted with valve stem oil seals which can be replaced if oil leakage occurs.

Valve Seat

The valve seat in an internal combustion gasoline or diesel engine is the surface against which an intake or an exhaust valve rests during the portion of the engine operating cycle when that valve is closed. The valve seat is a critical component of an engine in that if it is improperly positioned, oriented, or formed during manufacture, valve leakage will occur which will adversely affect the engine compression ratio and therefore the engine efficiency, performance (horsepower), exhaust emissions, and engine life.

Valve seats are often formed by first press-fitting an approximately cylindrical piece of a hardened metal alloy, such as Stellite, into a cast depression in a cylinder head above each eventual valve stem position, and then machining a conical-section surface into the valve seat that will mate with a corresponding conical-section of the corresponding valve. Generally two conical-section surfaces, one with a wider cone angle and one with a narrower cone-angle, are machined above and below the actual mating surface, to form the mating surface to the proper width (called "narrowing" the seat), and to enable it to be properly located with respect to the (wider) mating surface of the valve, so as to provide good sealing and heat transfer, when the valve is closed, and to provide good gas-flow characteristics through the valve, when it is opened.

Inexpensive engines may have valve seats that are simply cut into the material of the cylinder head or engine block (depending on the design of the engine). Some newer engines have seats that are sprayed on rather than being pressed into the head, allowing them to be thinner, creating more efficient transfer of heat through the valve seats, and enabling the valve stems to function at a lower temperature, thus allowing the valve stems (and other parts of the valvetrain) to be thinner and lighter.

There are several ways in which a valve seat may be improperly positioned or machined. These include incomplete seating during the press fitting-step, distortion of the nominally circular valve seat surfaces such they deviate unacceptably from perfect roundness or waviness, tilt of the machined surfaces relative to the valve guide hole axis, deviation of the valve seat surfaces from concentricity with the valve guide holes, and deviation of the machined conical section of the valve seat from the cone angle that is required to match the valve surface. Automated quality control of inserted and machined valve seats has traditionally been very difficult to achieve until the advent of digital holography which has enabled high-definition metrology for measuring all of these listed deviations.

Valve Shim

A valve shim is a disc of hardened metal of precisely calibrated height used to adjust the clearance (US:lash) between the back side of the cam-lobe and the valve on which it operates where the cam-lobe operates directly on the valve without an intermediate rocker arm. The shim sits atop a shim bucket which itself sits atop the valve-stem and moves up and down within a machined bore. While hydraulically self adjusting valves are sometimes used for their lack of required adjustment they do not work well in engines capable of high RPMs. In situations like these shims are most often used. As an example most motorcycle engines capable of 8000RPM+ use shim adjustable valves.

Valve Timing

In a piston engine, the valve timing is the precise timing of the opening and closing of the valves.

In four-stroke cycle engines and some two-stroke cycle engines, the valve timing is controlled by the camshaft. It can be varied by modifying the camshaft, or it can be varied during engine operation by the relatively new technology of variable valve timing. It is also affected by the adjustment of the valve mechanism, and particularly by the tappet clearance. However, this variation is normally unwanted.

Many two-stroke cycle and all wankel engines do not have a camshaft or valves, and the port timing can only be varied by machining the ports, and/or modifying the piston skirt (two stroke applications). Some supercharged two-stroke diesel engines do however have a cylinder head and camshaft similar to a four-stroke cycle engine.

In a steam engine, the control of the valve timing is an important part of the operation of the engine.

Trip Valve Gear

Trip valve mechanisms are a class of steam engine valve gear developed to improve efficiency. The trip mechanism allows the inlet valve to be closed rapidly, giving a short, sharp cut-off. The valve itself can be a drop valve or a Corliss valve.

Trip valve gear was applied to larger stationary engines. It was not used in transport applications, as it was not suitable for high speed.

The trip point of the valve mechanism, and therefore the cut-off, would be adjusted either manually or automatically by the governor. The valve is opened by the mechanical valve gear mechanism, and when the trip gear trigger releases the mechanism the valve is snapped closed, usually by a spring acting against a dashpot.

Two-stroke Power Valve System

The two-stroke power valve system is an improvement to a conventional two-stroke engine that gives a high power output over a wider RPM range.

Operation of a Two-stroke Engine

A stroke is the action of a piston travelling the full length of its cylinder. In a two-stroke engine, one of the two strokes combines the intake stroke and the compression stroke, while the other stroke combines the combustion stroke and the exhaust stroke.

As the piston travels upward in the cylinder, it creates low pressure area in the crankcase; this draws fresh air and atomized fuel from the carburetor through a hole in the cylinder wall or directly into the crankcase. As the piston continues travelling upward, transfer ports and the exhaust ports are closed off, thus trapping the combustible mixture in the combustion chamber. As the piston reaches the top of the cylinder, the mixture in the cylinder is compressed to the point of ignition.

The second stroke begins once ignition has taken place. The power stroke begins after the air-fuel mixture is ignited. The burnt fuel creates pressure in the cylinder above the piston and forces it downward. As the piston passes the midpoint of the downstroke, the exhaust port to the side of the cylinder is uncovered and initiates the flow of burned fuel out into the expansion chamber or muffler through the manifold. As the piston moves downward, where the air-fuel mixture remains from the previous intake-compression stroke. Shortly after the exhaust port is uncovered by the downward travel of the piston, the transfer ports begin to be uncovered. The transfer ports act as a passage through which the air-fuel mixture moves from the crankcase into the cylinder above the piston. Once the piston reaches the bottom of the stroke, the second cycle is completed and the process is repeated.

Engineering Design Improvements

The only moving parts inside simple two-stroke engines are the crankshaft, the connecting rod, and the piston. It is the same simplicity in design, however, that makes a two-stroke engine less fuel-efficient. At the bottom of the power stroke, the transfer ports, which deliver fresh fuel, are open at the same time as the exhaust port. This allows a significant amount of fresh fuel to run straight through the engine without ever being available for power production. Properly designed exhaust systems help minimise the amount of raw fuel loss in the exhaust process, but a two-stroke engine will always waste some fuel.

Many producers of two-stroke performance bikes fit them with the exhaust valve systems. These valves act to vary the height (and width) of the exhaust port thereby broadening power delivery over a wider rev range. Exhaust ports with fixed dimensions only produce usable power in a narrow rev range, which also affects fuel consumption and emissions.

In a race bike, this is not a problem as the engine will be operating at high RPM almost all the time. However, in a road/commuter bike, the limited power range is a problem. To provide more low RPM power, as well as enable the engine to produce a lot of high RPM power, a power valve system is used.

All power valve systems vary the duration of the exhaust port open time, which gives the engine usable low end power combined with excellent top end power. Manufacturers have also included sub exhaust chambers that extend the 'tuned length' of the expansion chamber.

Power valve actuation can be by mechanical (RPM dependent) or electric (servo motor) means increasingly with electronic control. Electronic control offers a greater degree of accuracy as well as being able to vary the opening of the valve and be tuned to conditions.

Suzuki AETC and Super AETC

AETC and Super AETC Suzuki engines, Automatic Exhaust Timing Control: The two-blade version was fitted to the VJ21 RGV250, and the three-blade version, to the VJ22 RGV250 and Suzuki RG150.

With the AETC system, the power-valve systems are normally partially closed at low RPM; when closed, it enables the engine to make more power. Up to a certain point, however, power drops off as the engine is unable to expel enough gases out of the exhaust. When the power-valve is opened, it allows more gases to flow out of the exhaust port. This system is recognisable by a small box above the exhaust outlet; the power-valves are situated in this box. Depending on the valve, they may be made of two (older version) or three (newer version) separate blades.

YPVS-Yamaha Power Valve System

YPVS Yamaha engines, Yamaha Power Valve System: Yamaha engineers realised that by altering the height of the exhaust port they could effectively change the engine power delivery thereby having optimal power and torque across the entire rev range, so it was that the YPVS was born. The valve is of a cylindrical design running across the top of the exhaust port, it is turned by a servo motor controlled from a control box taking information from the CDI (and other locations). The valve is a slightly ovular shape. This changes the height and size of the exhaust port at different engine speeds, maximising the available power at all rev ranges. Opening up firstly at 3k rpm for low end power, gradually in between 3-6k, fully opening at 6k rpm for maximum power, on most 125cc. It was fitted to all of the later models of the RZ/RD two-stroke road bikes (125, 250 350 and 500 cc), the TZR range. It was also added to the DT(125lc 2/3) range after 1984 (but was locked closed to comply with UK learner regulations until the(R) in 1988-04 which had a fully functional YPVS valve) the dt125r has a better design of engine, although not much altered in speed, just more reliable than its predecessor. The YZ series of motocross bikes has a mechanical power valve which is activated at RPM speed. The YPVS is only found on the liquid-cooled bikes not air cooled versions. Yamaha have also used a guillotine version in some of their later models such as the TZR250R SPR model..

Yamaha was actually the first company to produce consistent results with their YPVS in their race bikes. The 1977 OW35K was the first race bike to incorporate the power valve system and it won the Finnish GP in 1977. The Kadenacy effect was harnessed and controlled to a point that gave Yamaha great advantage over all the other manufacturers throughout the late 70's and into the mid 80's. The first street bikes with YPVS were the RZ/RD350 YPVS (LC2-onwards), and RZ/RD500 GP Replica in 1983-84.

Honda ATAC

ATAC System: The Honda Automatic Torque Amplification Chamber system works by effectively increasing or decreasing the volume of the exhaust system with a small butterfly valve located just before the exhaust connection. At low RPM a centrifugal crankshaft driven gear opens the valve into a small chamber and increases the volume of the exhaust by allowing the exhaust gases to flow through the chamber. At high RPM the ATAC valve is closed and the exhaust simply exits into the expansion chamber.

A larger expansion chamber allows for more power at lower RPM's because of the extra time needed for the impulse to "bounce" back for the supercharger effect. It was used on their CR motocrossers, GP bikes and MVX, NS and NSR road bikes.

Honda Power Port Valve

HPP valve. A centrifugal governor opens and closes a two-blade exhaust valve (using over 50 parts).

Honda V-TACS

The "V-TACS" - Variable Torque Amplification Chamber System - works differently from the "AETC system" and it will only work when it is used in conjunction with a tuned muffler. Tuned mufflers/expansion chambers increase power but only at the RPM they are designed for and can actually cause a power loss outside their tuned RPM. "V-TACS system" takes advantage of using an expansion chamber without losing power outside the expansion chamber's tuned RPM. Within the head and cylinder of the engine, there is a chamber that is sealed by a valve.

This sealed chamber is vented onto the exhaust port when the valve is open. At low RPM this valve is open, this has the effect of increasing the exhaust manifold volume and negating the power loss that would normally be apparent at low RPM with an expansion chamber. At mid RPM the valve is closed, this enables the expansion chamber to work. It is identified by the head and cylinder, being much larger than normal for its displacement, the cylinder wall is also cast with the wording VTACS on it.

V-TACS was a foot-operated power valve system made by Honda on some of its small two-stroke bikes and scooters, like the Honda FC50.

Honda RC-Valve

The Honda Revolution Control valve is designed and works in principle like the "AETC system." A small computer monitors engine RPM and adjusts a two-blade exhaust valve with an electric servo. Honda equipped many two-stroke motorcycles such as the NSR125 and NSR250 models with RC - Valve power plants.

Kawasaki KIPS

Kawasaki uses a power-valve system called KIPS (Kawasaki Integrated Powervalve System) on their two-stroke bikes. The KIPS utilises both alterations in port height, closing of the secondary port ducting, and a resonant chamber. KIPS is operated by a mechanical governor.

Crankcase Ventilation System

A crankcase ventilation system is a way for gases to escape in a controlled manner from the crankcase of an internal combustion engine. A common type of such system is a positive crankcase ventilation (PCV) system, the heart of which is a PCV valve—a variable-restriction valve that can react to changing pressure values and intermittently allow the passage of the gases to their intended destination (which nowadays is the engine's intake stream).

Internal combustion inevitably involves a small but continual amount of blow-by, which occurs when some of the gases from the combustion leak past the piston rings (that is, blow by them) to end up inside the crankcase. The gases could be vented through a simple hole or tube directly to the atmosphere, or they could "find their own way out" past baffles or past the oil seals of shafts or the gaskets of bolted joints. This is not a problem from a mechanical engineering viewpoint alone; but from other viewpoints, such as cleanliness for the user and environmental protection, such simple ventilation methods are not enough; escape of oil and gases must be prevented via a closed system that routes the escaping gases to the engine's intake stream and allows fresh air to come in.

History of Gradual Development

From the late 19th century through the early 20th, blow-by gases were allowed to find their own way out past seals and gaskets. It was considered normal for oil to be found both inside and outside an engine, and for oil to drip to the ground in small but constant amounts. This had also been true for steam engines and steam locomotives in the decades before. Even bearing and valve designs generally made little to no provision for keeping oil or waste gases contained. Sealed bearings and valve covers were for special applications only. For example, oilers kept the locomotives and rolling stock of railroads continually supplied with oil both inside and out. Although it was applied sparingly to oil cups and oil holes, it was not expected to stay hermetically sealed off from dripping and leaking to the wider environment. Gaskets and shaft seals were meant to limit loss of oil, but they were usually not expected to entirely prevent it. On internal combustion engines, the hydrocarbon-rich blow-by gases would diffuse through the oil in the seals and gaskets into the atmosphere. Engines with high amounts of blow-by (e.g., worn out ones, or ones not well built to begin with) would leak profusely via those routes.

From 1928 until the early 1960s, car and truck petrol engines vented combustion gases directly to the atmosphere through a simple vent tube. Frequently, this consisted of a pipe (the 'road draft tube') that extended out from the crankcase down to the bottom of the engine compartment. The bottom of the pipe was open to the atmosphere, and was placed such that when the car was in motion a slight vacuum was obtained, helping to extract combustion gases as they collected in the crankcase. The vacuum was satisfied by a vent, typically in the valve or valley cover, creating a constant flow of clean air through the engine's air volume. The oil mist would also be discharged, resulting in an oily film being deposited in the middle of each travel lane on heavily-used roads. The system was not positive though, as gases could travel both ways, or not move at all, depending on conditions. (Most modern diesel engines still use this type of system to dispose of crankcase fumes.)

During World War II, however, a different type of crankcase ventilation had to be invented to allow tank engines to operate during deep fording operations, where the normal draft tube ventilator would have allowed water to enter the crankcase and destroy the engine. The PCV system and its control valve were invented to meet this need, but no need for it on automobiles was recognised.

In 1952, Professor A. J. Haagen-Smit, of the California Institute of Technology at Pasadena, postulated that unburned hydrocarbons were a primary constituent of smog, and that gasoline powered automobiles were a major source of those hydrocarbons. After some investigation by the GM Research Laboratory (led by Dr. Lloyd L. Withrow), it was discovered in 1958 that the road draft tube was a major source, about half, of the hydrocarbons coming from the automobile. GM's Cadillac Division, which had built many tanks during WWII, recognised that the simple PCV valve could be used to become the first major reduction in automotive hydrocarbon emissions. After confirming the PCV valves' effectiveness at hydrocarbon reduction, GM offered the PCV solution to the entire U.S. automobile industry, royalty free, through its trade association, the Automobile Manufacturers Association (AMA). In the absence of any legislated requirement, the AMA members agreed to put it on all California cars voluntarily in 1961, with national application following one year later, in 1962.

In 1967, several years after its introduction into production, the PCV system became the subject of a U.S. federal grand jury investigation, when it was alleged by some industry critics that the AMA was conspiring to keep several such smog reduction devices on the shelf to delay further smog control. After eighteen months of investigation by U.S. Attorney Samuel Flatow, the grand jury returned a "no-bill" decision, clearing the AMA, but resulting in a "Consent Decree" that all U.S. automobile companies agreed not to work jointly on smog control activities for a period of ten years.

In the decades since, the legislation and regulation around emissions has tightened substantially, and the emissions of cars and light trucks have decreased substantially. Today's petrol engines continue to use PCV systems in addition to many other emissions control measures, many of which reduce toxic exhaust emissions by relying on sensors (to collect status information for computer input), engine control units (for information processing), and actuators (to translate computer output to the changing of conditions).

System Components and Operation Details

The PCV valve is only one part of the PCV system, which is essentially a variable and calibrated air leak, whereby the engine returns its crankcase combustion gases. Instead of the gases being vented to the atmosphere, gases are fed back into the intake manifold, to re-enter the combustion chamber as part of a fresh charge of air and fuel. The PCV system is not a classical "vacuum leak". All the air collected by the air cleaner (and metred by the mass flow sensor, on a fuel injected engine) goes through the intake manifold. The PCV system just diverts a small percentage of this air via the breather to the crankcase before allowing it to be drawn back in to the intake tract again. It is an "open system" in that fresh exterior air is continuously used to flush contaminants from the crankcase and into the combustion chamber.

The system relies on the fact that, while the engine is running under light load and moderate throttle opening, the intake manifold's air pressure is always less than crankcase air pressure. The lower pressure of the intake manifold draws air towards it, pulling air from the breather through the crankcase (where it dilutes and mixes with combustion gases), through the PCV valve, and into the intake manifold.

The PCV system usually consists of the breather tube and the PCV valve. The breather tube connects the crankcase to a clean source of fresh air—the air cleaner body. Usually, clean air from the air cleaner flows into this tube and into the engine after passing through a screen, baffle, or other simple system to arrest a flame front, to prevent a potentially explosive atmosphere within the engine crank case from being ignited from a back-fire in to the intake manifold. The baffle, filter, or screen also traps oil mist, and keeps it inside the engine.

Once inside the engine, the air circulates around the interior of the engine, picking up and clearing away combustion byproduct gases, including a large amount of water vapour which includes dissolved chemical combustion byproducts, then exits through another simple baffle, screen, or mesh to trap oil droplets before being drawn out through the PCV valve, and into the intake manifold. On some PCV systems, this oil baffling takes place in a discrete replaceable part called the oil separator.

During the mid 1960s, substantial work was completed on an entirely independent crankcase ventilation system. The *Engine Ventilation System* had its own air intake filter, a sizable crankcase gases filter, condensate chamber, and highly engineered air flow valve. The system recycles clean water vapour, filters light oil, and filters air into the intake system before the carburetor, resulting in lower carbon monoxide and hydrocarbon emissions and extended engine oil life. Ford Motor Company made this system a requirement on all its material handling equipment (lift trucks) in 1971. This system was also used extensively on over-the-road diesel trucks and irrigation pumps. The AMA's choice of catalytic converter made automotive use unlikely.

The PCV valve connects the crankcase to the intake manifold from a location more-or-less opposite the breather connection. Typical locations include the opposite valve cover that the breather tube connects to on a V engine. A typical location is the valve cover(s), although some engines place the valve in locations far from the valve cover. The valve is simple, but actually performs a complicated control function. An internal restrictor (generally a cone or ball) is held in "normal" (engine off, zero vacuum) position with a light spring, exposing the full size of the PCV opening to the intake manifold. With the engine running, the tapered end of the cone is drawn towards the opening in the PCV valve by manifold vacuum, restricting the opening proportionate to the level of engine vacuum vs. spring tension. At idle, the intake manifold vacuum is near maximum. It is at this time the least amount of blow by is actually occurring, so the PCV valve provides the largest amount of (but not complete) restriction. As engine load increases, vacuum on the valve decreases proportionally and blow by increases proportionally. With a lower level of vacuum, the spring returns the cone to the "open" position to allow more air flow. At full throttle, vacuum is much reduced, down to between 1.5 and 3" Hg. At this point the PCV valve is nearly useless, and most combustion gases escape via the "breather tube" where they are then drawn in to the engine's intake manifold anyway.

Should the intake manifold's pressure be *higher* than that of the crankcase (which can happen in a turbocharged engine, or under certain conditions, such as an intake backfire), the PCV valve closes to prevent reversal of the exhausted air back into the crankcase again. In many cases PCV valves were only used for a few years, the function being taken over by a port on constant depression carburetors such as the SU. This has no moving parts or diaphragm to jam, block or rip like many PCV valves. It also doesn't have a 'one-way' function but the lack of it was never a problem in intake backfire.

It is critical that the parts of the PCV system be kept clean and open, otherwise air flow will be insufficient. A plugged or malfunctioning PCV system will eventually damage an engine. PCV problems are primarily due to neglect or poor maintenance, typically engine oil change intervals that are inadequate for the engine's driving conditions. A poorly-maintained engine's PCV system will eventually become contaminated with sludge, causing serious problems. If the engine's lubricating oil is changed with adequate frequency, the PCV system will remain clear practically for the life of the engine. However, since the valve is operating continuously as one operates the vehicle, it will fail over time. Typical maintenance schedules for gasoline engines include PCV valve replacement whenever the air filter or spark plugs are replaced. The long life of the valve despite the harsh operating environment is due to the trace amount of oil droplets suspended in the air that flows through the valve that keep it lubricated.

Alternatives in Non-car-and-light-truck Usage

Not all petrol engines have PCV valves. Engines not subject to emission controls, such as certain off-road engines, retain road draft tubes. Dragsters use a scavenger system and venturi tube in the exhaust to draw out combustion gases and maintain a small amount of vacuum in the crankcase to prevent oil leaks on to the race track.

Small gasoline two stroke engines use the crankcase to partially compress incoming air. All blow by in these engines is burned in the regular flow of air and fuel through the engine. Many small four-cycle engines such as lawn mower engines and small gasoline generators, simply use a draft tube connected to the intake, between the air filter and carburetor, to route all blow by back into the intake mixture. The higher operating temperature of these small engines has a side effect of preventing large amounts of water vapour and light hydrocarbons from condensing in the engine oil.

Air Suction Valve

An air suction valve (SAV) is a subsystem used to reduce emissions in the exhaust gases of internal combustion engines.

Working Principle

When an engine exhaust valve is open, the pressure in the exhaust manifold is higher than atmospheric pressure. The exhaust manifold is configured in such a way that positive and negative pulses are produced during the operation of the cycle by designing it as a diffuser.

The SAV has three openings: one is connected to the manifold vacuum, another is connected to the exhaust pipe and the third is open to the atmosphere. When a negative pulse is induced in the exhaust, air from the atmosphere enters the exhaust manifold through a one-way reed valve and serves to oxidize the carbon monoxide and hydrocarbons in the exhaust.

Air Cut Valve System

This system consists of a spring-loaded diaphragm, and a shaft which is attached to the bottom. A manifold vacuum is applied over the diaphragm where the shaft end is towards the opening of the atmosphere.

During deceleration, the engine vacuum rises and acts against the spring to push the shaft downwards, thereby blocking the opening through which atmospheric air is passed to the exhaust.

Though there is a negative pulse during deceleration, the manifold vacuum blocks and cuts off the air to the exhaust manifold.

Used on the Honda Falcon NX4, mainly responsible for backfire or popping during deceleration when an open exhaust is added to the bike and if extra accessories is on vacuum circuit (Scottoiler, etc.).

This system is mainly used in Indian two wheeler which are manufactured by Hero Honda. The advantage of this system is low cost.

Pulsed Air Valve

The Royal Enfield Bullet, from year 2006-current, uses an Air Suction Valve to reduce emissions in its exhaust gases. Royal Enfield Motors calls their Air Suction Valve the PAV, or Pulsed Air Valve. It's function is the same, using a vacuum at the intake to trigger a valve that pulses atmospheric air into the exhaust. The original "iron barrel" style engine did this with a hose from the PAV going to the exhaust pipe, the newer (2003–2009) AVL aluminium engine pulses the gas directly into the exhaust port on the head of the engine itself.

Air Gap (plumbing)

An air gap, as it relates to the plumbing trade, is the unobstructed vertical space between the water outlet and the flood level of a fixture.

A simple example is the space between a wall mounted faucet and the sink rim (this space is the air gap). Water can easily flow from the faucet into the sink, but there is no way that water can flow from the sink into the faucet without modifying the system. This arrangement will prevent any contaminants in the sink from flowing into the potable water system by siphonage and is the least expensive form of backflow prevention.

A common use of the term "air gap" in home plumbing refers to a fixture that provides back-flow prevention for an installed dishwasher. This "air gap" is seen above the countertop as a small cylindrical fixture mounted parallel with the faucet. Below the countertop, the drain pipe of the dishwasher feeds the "top" of the air gap, and the "bottom" of the air gap is plumbed into the sink drain below the basket. When installed and maintained properly, the air gap works as described above, and prevents drain water from the sink from backing up into the dishwasher, possibly contaminating dishes. Water flowing from the fixture into the sink indicates a need for maintenance or repair.

To further illustrate the air gap, consider what could happen if the air gap were eliminated by attaching a hose to the faucet and lowering the hose into a sink full of contaminated water. Under the right conditions (if the water supply loses pressure and the sink is higher than the point at which the water supply enters the house, for instance), the dirty water in the sink will be siphoned into the water pipes through the hose and faucet. The dirty water then will be dispersed throughout the drinking water system.

All plumbing codes require backflow prevention in several ways. The plumbing fixture manufacturers build the fixtures to meet these codes. A plumber must not build cross-connections in his daily work practices, and Plumbing Inspectors look for improper designs or connections of piping and plumbing fixtures. A common misconception is that a "high loop" (routing a drain line above a sink's flood level, for instance) will provide the same function as an air gap; this is not true, because the continuous connection in such a case still will allow backflow through siphoning.

An air gap must meet the requirements of being two times the effective inner diametre of the pipe (2*D) in order to be sufficient.

Atmospheric Vacuum Breaker

An Atmospheric Vacuum Breaker (AVB) is a backflow prevention device used in plumbing to prevent backflow of non-potable liquids into the drinking water system. It is usually constructed of brass and resembles a 90-degree elbow with a hood on its top to allow air to enter the water system if a siphon attempts to form. Inside this elbow is a poppet valve that is held "up" by the water pressure found in the system, closing the air entrance to the device. If the pressure in the "upstream side" is reduced to atmospheric pressure or below, the poppet valve drops and allows air to enter the system, breaking the siphon.

These devices, since they work on atmospheric principles, cannot be installed in an enclosure containing air contaminants. As air is a fluid, those contaminants could be drawn into the device, thus fouling the pipes. AVBs must be installed a minimum of 6 inches above the highest usage point in the system (i.e. sprinkler, drip emitter, etc.) Underground installation of AVBs is *entirely ineffectual* in providing backflow protection as groundwater in the underground vault could be drawn into the water system, contaminating it.

The AVB should not be used for high hazard applications and must not be used with continuous pressure on the device, as the poppet would likely stick and the AVB would no longer function properly. A shutoff valve should never be placed downstream of any AVB, as this would result in continuous pressure on the AVB. The AVB is not a testable device.

Hydrostatic Loop

A hydrostatic loop, though not often used in plumbing practice, is an arrangement of pipes formed into a vertical loop to prevent backflow of water within the plumbing potable water system. Since a siphon has a maximum height that it can work (about 33 feet), a hydrostatic loop is built higher than 33 feet. There are several ways to prevent siphonage and an undesirable backflow of the water in a plumbing system.

Kingston Valve

A Kingston valve is a type of valve fitted in the bottom of a ship's fuel, water and ballast tanks. Named after its inventor John Kingston (1786-1847), an English engineer.

The valve connects the tank to the sea and when open, allows sea water to enter the tank for the purposes of cleaning the tanks and to admit water ballast.

On submarines Kingston valves are fitted at the bottom of the submarine's ballast tanks in order to admit water when the submarine dives. The valves allow water to enter the ballast tanks, the enclosed air escaping through the open main vents at the top of the ballast tanks. During peacetime the Kingston valves are closed when the submarine is on the surface, the valves and vents being opened to dive, but to reduce the time required for diving in wartime the Kingston valves are left permanently open when at sea, the water being kept out of the ballast tanks by the air pressure of the trapped air. This pressure is released when the vents are opened for diving, allowing the water to enter through the open Kingston valves. The above describes certain ballast tanks, such as safety, negative and bow buoyancy tanks on vintage submarines. In general, *main ballast tanks* have the Kingston valve atop the tank. This Kingston valve, controllable both manually and hydraulically, is known by some as *main vent operating gear.* Main ballast tanks are in pairs; one on each side of the boat. One Kingston valve serves a pair, but each tank has a vent riser, with air connections and stop valves in the vent riser. The tank bottom is open to the sea through flood ports. The trapped air in the tank provides positive buoyancy while on the surface, although sea water sloshing in through the flood ports must be removed occasionally from the ballast tank(s) when cruising on the surface by means of low pressure air. The tank is capped by a vent riser, a pipe coming out of the top of the tank that terminates at a Kingston valve. In the vent riser is a valve that when shut, prevents flooding the tank when the submarine is rigged for surface. When the sub is rigged for dive, this valve is opened and flooding the tank becomes an option. Normally, the tank is flooded when the Kingston valve is opened, usually hydraulically from the control room. This Kingston valve then is left open until time to surface comes. This is because a slight leak of air from the air line provided to blow the tank into the vent riser could slowly make the boat more buoyant. When surfacing, the Kingston valve gets shut so the introduced air gets trapped in the tank and pushes the water out the tank's bottom through the flood port. The Kingston valve associated with each ballast tank can be operated manually. The vent riser valves get opened or shut manually during the rig for surface or rig for dive procedure.

Kingston valves are sometimes used to scuttle or sink a ship, as was done for example, to the German aircraft carrier Graf Zeppelin.

Pressure Balanced Valve

A pressure balanced valve, required by building codes in some municipalities, provides water at nearly constant temperature to a shower or bathtub, despite pressure fluctuations in either the hot or cold supply lines.

If, for example, someone flushes a toilet while the shower is in use, the fixture suddenly draws a significant amount of cold water from the common supply line, causing a pressure drop. In the absence of a compensating mechanism, the relatively higher pressure in the hot water supply line will cause the shower temperature to rise just as suddenly, possibly reaching an uncomfortable or even dangerous level. Conversely, if someone opens a hot water faucet elsewhere, the relatively higher pressure in the cold water supply line will cause the shower temperature to drop suddenly. The pressure-balanced shower valve compensates for changes in water pressure. It has a diaphragm or piston inside that reacts with a relative changes in either hot or cold water pressure to maintain balanced pressure. As water pressure drops on one supply line, the valve reduces the pressure in the other supply line to match. A side effect of this is that the pressure and flow at the shower head or tub spigot will drop twice as much as if only one supply line had been affected, but without a large temperature change. There are ball bearings in the valves to regulate forces.

Municipalities that require the use of pressure balanced valves do so to prevent scalding injuries, in particular to the elderly, the infirm, to children and infants.

Blowout preventer

A blowout preventer is a large, specialised valve used to seal, control and monitor oil and gas wells. Blowout preventers were developed to cope with extreme erratic pressures and uncontrolled flow (formation kick) emanating from a well reservoir during drilling. Kicks can lead to a potentially catastrophic event known as a blowout. In addition to controlling the downhole (occurring in the drilled hole) pressure and the flow of oil and gas, blowout preventers are intended to prevent tubing (e.g. drill pipe and well casing), tools and drilling fluid from being blown out of the wellbore (also known as bore hole, the hole leading to the reservoir) when a blowout threatens. Blowout preventers are critical to the safety of crew, rig (the equipment system used to drill a wellbore) and environment, and to the monitoring and maintenance of well integrity; thus blowout preventers are intended to be fail-safe devices.

The term BOP (an initialism rather than a spoken acronym, i.e., pronounced B-O-P, not "bop") is used in oilfield vernacular to refer to blowout preventers.

The abbreviated term preventer, usually prefaced by a type (e.g. ram preventer), is used to refer to a single blowout preventer unit. A blowout preventer may also simply be referred to by its type (e.g. ram).

The terms blowout preventer, blowout preventer stack and blowout preventer system are commonly used interchangeably and in a general manner to describe an assembly of several stacked blowout preventers of varying type and function, as well as auxiliary components. A typical subsea deepwater blowout preventer system includes components such as electrical and hydraulic lines, control pods, hydraulic accumulators, test valve, kill and choke lines and valves, riser joint, hydraulic connectors, and a support frame.

Two categories of blowout preventer are most prevalent: *ram* and *annular*. BOP stacks frequently utilise both types, typically with at least one annular BOP stacked above several ram BOPs.

(A related valve, called an *inside blowout preventer*, *internal blowout preventer*, or *IBOP*, is positioned within, and restricts flow up, the drillpipe. This article does not address inside blowout preventer use.) Blowout preventers are used at land and offshore rigs, and subsea. Land and subsea BOPs are secured to the top of the wellbore, known as the wellhead. BOPs on offshore rigs are mounted below the rig deck. Subsea BOPs are connected to the offshore rig above by a drilling riser that provides a continuous pathway for the drill string and fluids emanating from the wellbore. In effect, a riser extends the wellbore to the rig.

Use

The invention of blowout preventers was instrumental in reducing the incidence of oil gushers, blowouts, indicating that substantial improvement is needed.

Blowout preventers come in a variety of styles, sizes and pressure ratings. Several individual units serving various functions are combined to compose a blowout preventer stack. Multiple blowout preventers of the same type are frequently provided for redundancy, an important factor in the effectiveness of fail-safe devices.

The primary functions of a blowout preventer system are to:

- Confine well fluid to the wellbore;
- Provide means to add fluid to the wellbore;
- Allow controlled volumes of fluid to be withdrawn from the wellbore.

Additionally, and in performing those primary functions, blowout preventer systems are used to:

- Regulate and monitor wellbore pressure;
- Centre and hang off the drill string in the wellbore;
- Shut in the well (e.g. seal the void, annulus, between drillpipe and casing);
- "Kill" the well (prevent the flow of formation fluid, influx, from the reservoir into the wellbore) ;
- Seal the wellhead (close off the wellbore);
- Sever the casing or drill pipe (in case of emergencies).

In drilling a typical high-pressure well, drill strings are routed through a blowout preventer stack toward the reservoir of oil and gas. As the well is drilled, drilling fluid, "mud", is fed through the drill string down to the drill bit, "blade", and returns up the wellbore in the ring-shaped void, annulus, between the outside of the drill pipe and the casing (piping that lines the wellbore). The column of drilling mud exerts downward hydrostatic pressure to counter opposing pressure from the formation being drilled, allowing drilling to proceed.

When a kick (influx of formation fluid) occurs, rig operators or automatic systems close the blowout preventer units, sealing the annulus to stop the flow of fluids out of the wellbore. Denser mud is then circulated into the wellbore down the drill string, up the annulus and out through the choke line at the base of the BOP stack through chokes (flow restrictors) until downhole pressure is overcome. Once "kill weight" mud extends from the bottom of the well to the top, the well has been "killed". If the integrity of the well is intact drilling may be resumed. Alternatively, if circulation is not feasible it may be possible to kill the well by "bullheading", forcibly pumping, in the heavier mud from the top through the kill line connection at the base of the stack. This is less desirable because of the higher surface pressures likely needed and the fact that much of the mud originally in the annulus must be forced into receptive formations in the open hole section beneath the deepest casing shoe.

If the blowout preventers and mud do not restrict the upward pressures of a kick, a blowout results, potentially shooting tubing, oil and gas up the wellbore, damaging the rig, and leaving well integrity in question.

Since BOPs are important for the safety of the crew and natural environment, as well as the drilling rig and the wellbore itself, authorities recommend, and regulations require, that BOPs be regularly inspected, tested and refurbished. Tests vary from daily test of functions on critical wells to monthly or less frequent testing on wells with low likelihood of control problems.

Exploitable reservoirs of oil and gas are increasingly rare and remote, leading to increased subsea deepwater well exploration and requiring BOPs to remain submerged for as long as a year in extreme conditions. As a result, BOP assemblies have grown larger and heavier (e.g. a single ram-type BOP unit can weigh in excess of 30,000 pounds), while the space allotted for BOP stacks on existing offshore rigs has not grown commensurately. Thus a key focus in the technological development of BOPs over the last two decades has been limiting their footprint and weight while simultaneously increasing safe operating capacity.

Types

BOPs come in two basic types, ram and annular. Both are often used together in drilling rig BOP stacks, typically with at least one annular BOP capping a stack of several ram BOPs.

Ram Blowout Preventer

The ram BOP was invented by James Smither Abercrombie and Harry S. Cameron in 1922, and was brought to market in 1924 by Cameron Iron Works.

A ram-type BOP is similar in operation to a gate valve, but uses a pair of opposing steel plungers, rams. The rams extend toward the centre of the wellbore to restrict flow or retract open in order to permit flow. The inner and top faces of the rams are fitted with packers (elastomeric seals) that press against each other, against the wellbore, and around tubing running through the wellbore. Outlets at the sides of the BOP housing (body) are used for connection to choke and kill lines or valves.

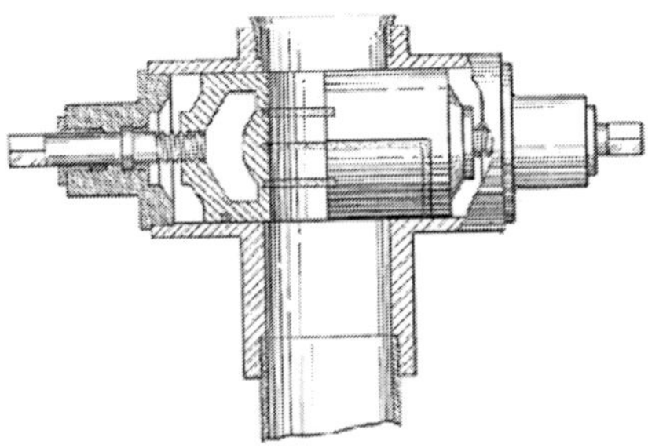

Figure: A Patent Drawing of the Original Ram-type Blowout Preventer, by Cameron Iron Works (1922).

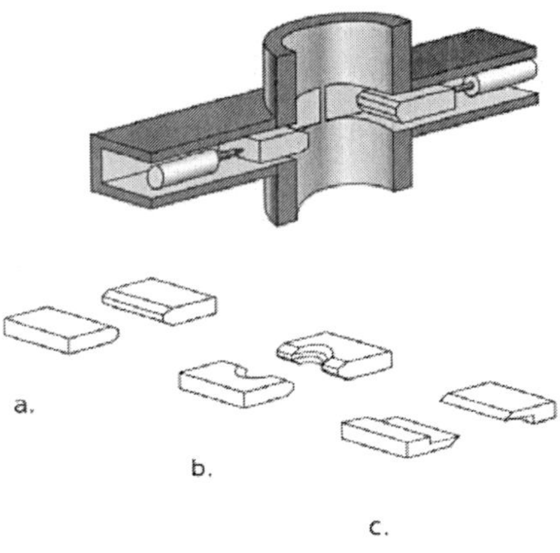

Figure: Blowout Preventer diagram showing different types of rams. (a) blind ram (b) pipe ram and (c) shear ram.

Rams, or ram blocks, are of four common types: *pipe*, *blind*, *shear*, and *blind shear*.

Pipe rams close around a drill pipe, restricting flow in the annulus (ring-shaped space between concentric objects) between the outside of the drill pipe and the wellbore, but do not obstruct flow within the drill pipe. Variable-bore pipe rams can accommodate tubing in a wider range of outside diametres than standard pipe rams, but typically with some loss of pressure capacity and longevity.

Blind rams (also known as sealing rams), which have no openings for tubing, can close off the well when the well does not contain a drill string or other tubing, and seal it.

Shear rams cut through the drill string or casing with hardened steel shears.

Blind shear rams (also known as shear seal rams, or sealing shear rams) are intended to seal a wellbore, even when the bore is occupied by a drill string, by cutting through the drill string as the rams close off the well. The upper portion of the severed drill string is freed from the ram, while the lower portion may be crimped and the "fish tail" captured to hang the drill string off the BOP.

In addition to the standard ram functions, variable-bore pipe rams are frequently used as test rams in a modified blowout preventer device known as a stack test valve. Stack test valves are positioned at the bottom of a BOP stack and resist downward pressure (unlike BOPs, which resist upward pressures). By closing the test ram and a BOP ram about the drillstring and pressurizing the annulus, the BOP is pressure-tested for proper function.

The original ram BOPs of the 1920s were simple and rugged manual devices with minimal parts. The BOP housing (body) had a vertical well bore and horizontal ram cavity (ram guide chamber). Opposing rams (plungers) in the ram cavity translated horizontally, actuated by threaded ram shafts (piston rods) in the manner of a screw jack. Torque from turning the ram shafts by wrench or hand wheel was converted to linear motion and the rams, coupled to the inner ends of the ram shafts, opened and closed the well bore. Such screw jack type operation provided enough mechanical advantage for rams to overcome downhole pressures and seal the wellbore annulus.

Hydraulic rams BOPs were in use by the 1940s. Hydraulically actuated blowout preventers had many potential advantages. The pressure could be equalized in the opposing hydraulic cylinders causing the rams to operate in unison. Relatively rapid actuation and remote control were facilitated, and hydraulic rams were well-suited to high pressure wells.

Because BOPs are fail-safe devices, efforts to minimise the complexity of the devices are still employed to ensure ram BOP reliability and longevity. As a result, despite the ever-increasing demands placed on them, state of the art ram BOPs are conceptually the same as the first effective models, and resemble those units in many ways.

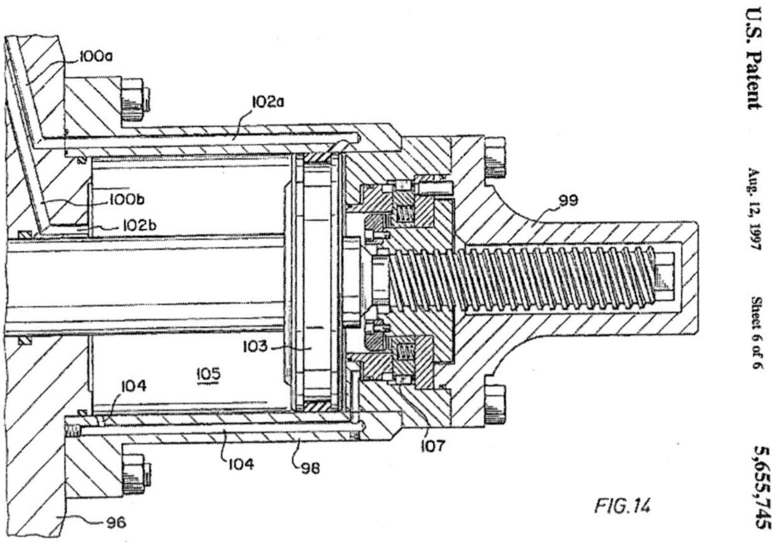

Figure: Hydril Company's Compact BOP Ram Actuator Assembly Patent Drawing

Ram BOPs for use in deepwater applications universally employ hydraulic actuation. Threaded shafts are often still incorporated into hydraulic ram BOPs as lock rods that hold the ram in position after hydraulic actuation. By using a mechanical ram locking mechanism, constant hydraulic pressure need not be maintained. Lock rods may be coupled to ram shafts or not, depending on manufacturer. Other types of ram locks, such as wedge locks, are also used.

Typical ram actuator assemblies (operator systems) are secured to the BOP housing by removable bonnets. Unbolting the bonnets from the housing allows BOP maintenance and facilitates the substitution of rams. In that way, for example, a pipe ram BOP can be converted to a blind shear ram BOP.

Shear-type ram BOPs require the greatest closing force in order to cut through tubing occupying the wellbore. Boosters (auxiliary hydraulic actuators) are frequently mounted to the outer ends of a BOP's hydraulic actuators to provide additional shearing force for shear rams. Ram BOPs are typically designed so that well pressure will help maintain the rams in their closed, sealing position. That is achieved by allowing fluid to pass through a channel in the ram and exert pressure at the ram's rear and toward the centre of the wellbore. Providing a channel in the ram also limits the thrust required to overcome well bore pressure. Single ram and double ram BOPs are commonly available. The names refer to the quantity of ram cavities (equivalent to the effective quantity of valves) contained in the unit. A double ram BOP is more compact and lighter than a stack of two single ram BOPs while providing the same functionality, and is thus desirable in many applications. Triple ram BOPs are also manufactured, but not as common.

Technological development of ram BOPs has been directed towards deeper and higher pressure wells, greater reliability, reduced maintenance, facilitated replacement of components, facilitated ROV intervention, reduced hydraulic fluid consumption, and improved connectors, packers, seals, locks and rams. In addition, limiting BOP weight and footprint are significant concerns to account for the limitations of existing rigs.

The highest-capacity large-bore ram blowout preventer on the market, as of July 2010, Cameron's EVO 20K BOP, has a hold-pressure rating of 20,000 psi, ram force in excess of 1,000,000 pounds, and a well bore diametre of 18.75 inches.

Annular blowout preventer

The annular blowout preventer was invented by Granville Sloan Knox in 1946; a U.S. patent for it was awarded in 1952. Often around the rig it is called the "Hydril", after the name of one of the manufacturers of such devices.

An annular-type blowout preventer can close around the drill string, casing or a non-cylindrical object, such as the kelly. Drill pipe including the larger-diametre tool joints (threaded connectors) can be "stripped" (i.e., moved vertically while pressure is contained below) through an annular preventer by careful control of the hydraulic closing pressure. Annular blowout preventers are also effective at maintaining a seal around the drillpipe even as it rotates during drilling. Regulations typically require that an annular preventer be able to completely close a wellbore, but annular preventers are generally not as effective as ram preventers in maintaining a seal on an open hole. Annular BOPs are typically located at the top of a BOP stack, with one or two annular preventers positioned above a series of several ram preventers.

An annular blowout preventer uses the principle of a wedge to shut in the wellbore. It has a donut-like rubber seal, known as an elastomeric packing unit, reinforced with steel ribs. The packing unit is situated in the BOP housing between the head and hydraulic piston.

When the piston is actuated, its upward thrust forces the packing unit to constrict, like a sphincter, sealing the annulus or openhole. Annular preventers have only two moving parts, piston and packing unit, making them simple and easy to maintain relative to ram preventers.

The original type of annular blowout preventer uses a "wedge-faced" (conical-faced) piston. As the piston rises, vertical movement of the packing unit is restricted by the head and the sloped face of the piston squeezes the packing unit inward, toward the centre of the wellbore.

In 1972, Ado N. Vujasinovic was awarded a patent for a variation on the annular preventer known as a spherical blowout preventer, so-named because of its spherical-faced head. As the piston rises the packing unit is thust upward against the curved head, which constricts the packing unit inward. Both types of annular preventer are in common use.

Control methods

When rigs are drilled on land or in very shallow water where the wellhead is above the water line, BOPs are activated by hydraulic pressure from a remote accumulator. Several control stations will be mounted around the rig. They also can be closed manually by turning large wheel-like handles. In deeper offshore operations with the wellhead just above the mudline on the sea floor, there are four primary ways by which a BOP can be controlled. The possible means are:

- Electrical Control Signal: sent from the surface through a control cable;
- Acoustical Control Signal: sent from the surface based on a modulated/encoded pulse of sound transmitted by an underwater transducer;
- ROV Intervention: remotely operated vehicles (ROVs) mechanically control valves and provide hydraulic pressure to the stack (via "hot stab" panels);
- Deadman Switch / Auto Shear: fail-safe activation of selected BOPs during an emergency, and if the control, power and hydraulic lines have been severed.

Two control pods are provided on the BOP for redundancy. Electrical signal control of the pods is primary. Acoustical, ROV intervention and dead-man controls are secondary. An emergency disconnect system, or EDS, disconnects the rig from the well in case of an emergency. The EDS is also intended to automatically trigger the deadman switch, which closes the BOP, kill and choke valves. The EDS may be a subsystem of the BOP stack's control pods or separate.

Pumps on the rig normally deliver pressure to the blowout preventer stack through hydraulic lines. Hydraulic accumulators are on the BOP stack enable closure of blowout preventers even if the BOP stack is disconnected from the rig. It is also possible to trigger the closing of BOPs automatically based on too high pressure or excessive flow.

Individual wells along the U.S. coastline may also be required to have BOPs with backup acoustic control. General requirements of other nations, including Brazil, were drawn to require this method. BOPs featuring this method may cost as much as US$500,000 more than those that omit the feature.

Deepwater Horizon Blowout

During the *Deepwater Horizon* drilling rig explosion incident on April 20, 2010, the blowout preventer should have been activated automatically, cutting the drillstring and sealing the well to preclude a blowout and subsequent oil spill in the Gulf of Mexico, but it failed to fully engage. Underwater robots (ROVs) later were used to manually trigger the blind shear ram preventer, to no avail.

As of May 2010 it was unknown why the blowout preventer failed. Chief surveyor John David Forsyth of the American Bureau of Shipping testified in hearings before the Joint Investigation of the Minerals Management Service and the U.S. Coast Guard investigating the causes of the explosion that his agency last inspected the rig's blowout preventer in 2005. BP representatives suggested that the preventer could have suffered a hydraulic leak. Gamma-ray imaging of the preventer conducted on May 12 and May 13, 2010 showed that the preventer's internal valves were partially closed and were restricting the flow of oil. Whether the valves closed automatically during the explosion or were shut manually by remotely operated vehicle work is unknown. A statement released by Congressman Bart Stupak revealed that, among other issues, the emergency disconnect system (EDS) did not function as intended and may have malfunctioned due to the explosion on the Deepwater Horizon.

The permit for the Macondo Prospect by the Minerals Management Service in 2009 did not require redundant acoustic control means. Inasmuch as the BOPs could not be closed successfully by underwater manipulation (ROV Intervention), pending results of a complete investigation, it is uncertain whether this omission was a factor in the blowout.

Documents discussed during congressional hearings June 17, 2010, suggested that a battery in the device's control pod was flat and that the rig's owner, Transocean, may have "modified" Cameron's equipment for the Macondo site (including incorrectly routing hydraulic pressure to a stack test valve instead of a pipe ram BOP) which increased the risk of BOP failure, in spite of warnings from their contractor to that effect. Another hypothesis is that a junction in the drilling pipe may have been positioned in the BOP stack in such way that its shear rams had an insurmountable thickness of material to cut through.

It was later discovered that a second piece of tubing got into the BOP stack at some point during the Macondo incident, potentially explaining the failure of the BOP shearing mechanism. As of July 2010 it is unknown whether the tubing might be casing that shot up through the well or perhaps broken drill pipe that dropped into the well.

On July 10, 2010 BP began operations to install a sealing cap, also known as a capping stack, atop the failed blowout preventer stack. Based on BP's video feeds of the operation the sealing cap assembly, called Top Hat 10, includes a stack of three blind shear ram BOPs manufactured by Hydril (a GE Oil & Gas company), one of Cameron's chief competitors. By July 15 the 3 ram capping stack had sealed the Macondo well, if only temporarily, for the first time in 87 days. The U.S. government wants the failed blowout preventer to be replaced in case of any pressure that occurs when the relief well intersects with the well. On September 3 at 1:20 p.m. CDT the 300 ton failed blowout preventer was removed from the well and began being slowly lifted to the surface. Later that day a replacement blowout preventer was placed on the well. On September 4 at 6:54 p.m. CDT the failed blowout preventer reached the surface of the water and at 9:16 p.m. CDT it was placed in a special container on board the vessel Helix Q4000. The failed blowout preventer was taken to a NASA facility in Louisiana for examination by Det Norske Veritas (DNV).

On 20 March 2011, DNV presented their report to the US Department of Energy. Their primary conclusion was that the rams failed to shear through the oil pipe and seal it because it has buckled out of the line of action of the rams. They did not suggest any failure of actuation as would be caused by faulty batteries.

Reduced Pressure Zone Device

A Reduced Pressure Zone Device (RPZD, RPZ or RPZ valve) is a type of backflow prevention device used to protect water supplies from contamination. RPZDs may also be known as *reduced pressure principle backflow prevention devices*, or *reduced pressure zone assemblies*.

Description

A RPZD is considered suitable for significant hazard applications, that is, where the consequence of backflow into the water supply would cause significant harm, although not for the highest risks, such as human waste. They are considered suitable because they prevent both backflow and back-siphonage, because of a redundant design (even with two check valves broken the device still provides protection), and because they are testable to verify correct operation.

RPZD are often chosen instead of an air gap since the head loss across the RPZ is much smaller.

The device consists of two independent check valves, plumbed in series, with a pressure monitored chamber between. The chamber is maintained at a pressure that is lower than the water supply pressure, but high enough to be useful downstream. The reduced pressure is guaranteed by a differential pressure relief valve, which automatically relieves excess pressure in the chamber by discharging to a drain. Four test cocks are installed to allow an inspector to verify proper operation of the valve with a differential pressure metre. Two gate valves (one before and one after the RPZ) are provided to allow for testing and repair. Facilities which require constant water pressure must install two RPZs in parallel so that one device will be available to supply the facility while the other is tested or repaired.

Discharge from the relief device is an indication that either of the two check valves is 'passing' (leaking past their shutoff seats) or the relief valve itself is faulty. Some normal water use conditions may cause short episodes of discharge. In the case of the upstream check valve passing, the differential pressure (higher supply pressure compared with the lower chamber pressure) causes any flow to occur only in one direction. A buildup of pressure in the chamber in such a case would be relieved to the drain.

In the case of the downstream check valve passing, the differential pressure relief valve prevents the possibility of the chamber pressure from exceeding the supply pressure.

It is theoretically possible that, when both valves are faulty, backflow could occur, should the reverse flow rate exceed the capacity of the relief valve.

Because certain combinations of check valve failure and/or system backpressure cause the relief valve to discharge, the device must be mounted in a location where the drain will not become flooded.

An example of where backflow would harm the water supply is the use of well washing devices inside underground sewerage pumping stations. At times untreated sewerage may contain a variety of harmful gases that will effectively break down and deteriorate concrete wells, hence well washers are utilised to spray water and wash down contaminated concrete walls of a well. All well washers are installed with RPZ Devices in case a pumping station breaks down, and the sewerage level rises above the well washer causing backflow down the water supply line.

A reduced pressure zone device is tested with a device called a differential gauge. This gauge uses hoses attached to the various test cocks of the RPZ and checks to be certain that the check valve springs as well as the relief valve spring are not fouled and are working properly. There are a variety of manufacturers of these gauges, but most utilise either a three valve or a five valve design to test the RPZ in different ways.

In the UK, New York, Texas, as well as St. Louis city and county, the RPZ valve must be tested at least every 12 months by an accredited tester.

In the UK all RPZ Valve Testers have to be WRAS Accredited

Sloan Valve Company

The company was founded by William Elvis Sloan in Chicago, Illinois. When he initially sold his Royal flushometre in 1906, he only sold one at the time because customers were wary of adjusting to his invention. Even some plumbing fixture manufacturers wouldn't sell their products if they were furnished with Sloan's flushometres. The next year (1907), he sold two units; in 1908 he sold 150 units, and the number of units sold have risen up since then.

Sloan's other flushometres besides the Royal included two piston-type flushometres, the Crown and the Gem. A special piston-type flushometre, called the Naval, was built for marine applications. In 1976 Sloan introduced the Optima line of sensor-activated flushometres (the first few were installed at Chicago's O'Hare International Airport), in which battery-powered equivalents under the name Optima Plus would follow in 1991. Manual flushometres are now available in numerous types and styles, such as dual-flush models, which enable users to choose from a regular 1.6 gallon flush or a reduced flush for greater water savings, and various water-saver models for water closets and urinals.

In addition to manually operated flushometres, Sloan also offers sensor-activated flushometres and other plumbing fixtures, including water-efficient electronic faucets, solid-surface lavatory systems, scrub sinks and wash stations, bedpan washer flushometres, Waterfree urinals, and a 1.0-gallon pressure-assist system for water closets. Other restroom fixtures include sensor-activated hand dryers, low-flow showerheads, networked water-control systems, and electronic soap dispensers.Steam trap

A steam trap is a device used to discharge condensate and non condensable gases with a negligible consumption or loss of live steam. Most steam traps are nothing more than automatic valves. They open, close or modulate automatically. Others, like venturi traps, are based on turbulent 2-phase flows to obstruct the steam flow.

The three important functions of steam traps are:

1. Discharge condensate as soon as it is formed.
2. Have a negligible steam consumption.
3. Have the capability of discharging air and other non-condensable gases.

Basic Operation

The best and simplest form of steam trap is a disc or short solid pipe nipple with a small hole drilled through it installed at the lowest point of the equipment. Since steam condensate will collect at the lowest point and live steam is as about 1200 greater in volume than this hot liquid, condensate is effectively removed and steam is blocked.

Mechanical steam traps basically open when condensate needs to be removed, and close when there is only steam present.

Types

Steam traps can be split into four major types:

1. Mechanical traps. They have a float that rises and falls in relation to condensate level and this usually has a mechanical linkage attached that opens and closes the valve. Mechanical traps operate in direct relationship to condensate levels present

in the body of the steam trap. Inverted bucket and float traps are examples of mechanical traps.

2. Temperature traps. They have a valve that is driven on / off the seat by either expansion / contraction caused by temperature change. They differ from mechanical traps in that their design requires them to hold back some condensate waiting for it to cool sufficiently to allow the valve to open. In most circumstances this is not desirable as condensate needs to be removed as soon as it is formed. Thermostatic traps, Bi-Thermostatic traps and bimetallic traps are examples of temperature operated traps.
3. Thermodynamic (TD) traps. Thermodynamic traps work on the difference in dynamic response to velocity change in flow of compressible and incompressible fluids. As steam enters, static pressure above the disk forces the disk against the valve seat. The static pressure over a large area overcomes the high inlet pressure of the steam. As the steam starts to condense, the pressure against the disk lessens and the trap cycles. This essentially makes a TD trap a "time cycle" device: it will open even if there is only steam present, this can cause premature wear. If non condensable gas is trapped on top of the disc, it can cause the trap to be locked shut.
4. Venturi (orifice) traps. This type works in a turbulent two-phase flow regime. Internally it consists of a venturi type valve with a certain shape. Condensate is fully discharged while eventual steam also tries to pass the venturi. But while traversing the venturi towards the low pressure zone the steam expands and chokes the throughput together with the slow condensate. Because of this, the amount of live steam escaping the orifice is negligible.

Tesla Valve

A Tesla Valve is a valve allowing for a fluid to flow very preferentially (or entirely) in one direction, but without the moving parts usually found in valves. The principle is that currents flow along different paths in different directions, and that these differences have a disproportionate effect on the resistance of the valve. Named after Nikola Tesla, it was invented in 1916 (U.S. patent 1,329,559; patented 1920). In practice, the idea has seen little development or use since its invention. A similar mechanism is used for mixing on a small scale through the use of a Coanda effect mixer.

Chapter 9

Vacuum Breaker

A vacuum breaker is an attachment commonly placed on a bibcock valve or toilet or urinal flush valve, that prevents water from being siphoned backward into the public drinking water system. This prevents contamination should the public drinking water system's pressure drop. A vacuum breaker typically contains a plastic disc that is pressed forward by water supply pressure, and covers small vent holes. Should the supply pressure drop, the disc springs back opening the vent holes (which let in outside air), and preventing backflow of water.

A more complex valve that accomplishes much the same purpose is the backflow preventer. Vacuum relief valves are sometimes known as vacuum breakers.

Corliss Steam Engine

A Corliss steam engine (or Corliss engine) is a steam engine, fitted with rotary valves and with variable valve timing patented in 1849, invented by and named after the American engineer George Henry Corliss in Providence, Rhode Island.

Engines fitted with Corliss valve gear offered the best thermal efficiency of any type of stationary steam engine until the refinement of the uniflow steam engine and steam turbine in the 20th century. Corliss engines were generally about 30 percent more fuel efficient than conventional steam engines with fixed cutoff. This increased efficiency made steam power more economical than water power, allowing industrial development away from millponds.

Corliss engines were typically used as stationary engines to provide mechanical power to line shafting in factories and mills and to drive dynamos to generate electricity. Many were quite large, standing several stories tall, and developing several hundred horsepower, albeit at low speed, turning massive flywheels weighing several tons at about 100 revolutions per minute. Some of these engines had unique roles as mechanical legacy systems and because of their relatively high efficiency and low maintenance requirements, some remain in service in early 21st century.

Corliss Engine Mechanisms

Corliss engines have four valves for each cylinder, with steam and exhaust valves located at each end. Corliss engines incorporate distinct refinements in both the valves themselves and in the valve gear, that is, the system of linkages that operate the valves.

The use of separate valves for admission and exhaust means that neither the valves nor the steam passages between cylinders and valves need to change temperature during the power and exhaust cycle, and it means that the timing of the admission and exhaust valves can be independently controlled. In contrast, conventional steam engines have a slide valve or piston valve that alternately feeds and exhausts through passages to each end of the cylinder. These passages are exposed to wide temperature swings during engine operation, and there are high temperature gradients within the valve mechanism itself.

Clark (1891) commented that the Corliss gear 'is essentially a combination of elements previously known and used separately, affecting the cylinder and the valve-gear'. The origins of the Corliss gear with regard to previous steam valve gear was traced by Inglis (1868).

Corliss Valve Gear

Figure: "Gordon's improved Corliss valvegear", detailed view. The wrist-plate is the central plate from which rods radiate to each of the 4 valves.

George Corliss received US Patent 6162 for his valve gear on March 10, 1849. This patent covered the use of a *wrist-plate* to convey the valve motion from a single eccentric to the four valves of the engine, and it covered the use of trip valves with variable cutoff under governor control that characterise Corliss Engines. Unlike later engines, most of which were horizontal, this patent describes a vertical cylinder beam engine, and it used individual slide valves for admission and exhaust at each end of the cylinder.

The inlet valves are pulled open with an eccentric-driven pawl; when the pawl trips, the rapid closure is damped using a dashpot. In many engines, the same dashpot acts as a vacuum spring to pull the valves closed, but Corliss's early engines were slow enough that it was the weight of the dashpot piston and rod that closed the valve.

The speed of a Corliss engine is controlled by varying the cutoff of steam during each power stroke, while leaving the throttle wide open at all times. To accomplish this, the centrifugal governor is linked to a pair of cams, one for each admission valve. These cams determine the point during the piston stroke that the pawl will release, allowing that valve to close.

As with all steam engines where the cutoff can be regulated, the virtue of doing so lies in the fact that most of the power stroke is powered by the expansion of steam in the cylinder after the admission valve has closed. This comes far closer to the ideal Carnot cycle than is possible with an engine where the admission valve is open for the length of the power stroke and speed is regulated by a throttle valve.

The Corliss valve gearing allowed more uniform speed and better response to load changes, making it suitable for applications like rolling mills and spinning, and greatly expanding its use in manufacturing.

Corliss Valves

Corliss valves open directly into the cylinder. The valves connect the cylinder to separate steam and exhaust plenums. Initially, Corliss used slide valves with linear actuators, but by 1851, Corliss had shifted to semi-rotary valve actuators, as documented in U.S. Patent 8253. In this engine, the wrist plate was moved to the centre of the cylinder side, as on later Corliss engines. This was still a beam engine, however, and the semi-rotary valve actuators operated linear slide valves inside the four valve chests of the engine. Corliss valves are in the form of a minor circular segment, rotating inside a cylindrical valve-face. Their actuating mechanism is off along the axis of the valve, thus they have little "dead space" such as the stem of a poppet valve and the entire port area can be used efficiently for gas flow.

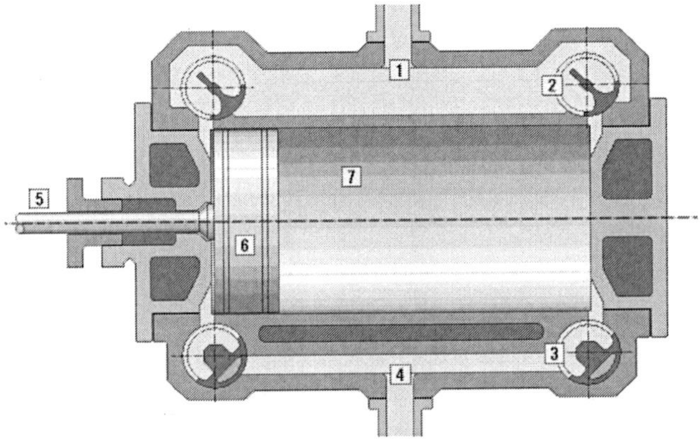

Figure: Sectioned cylinder, showing rotating valves

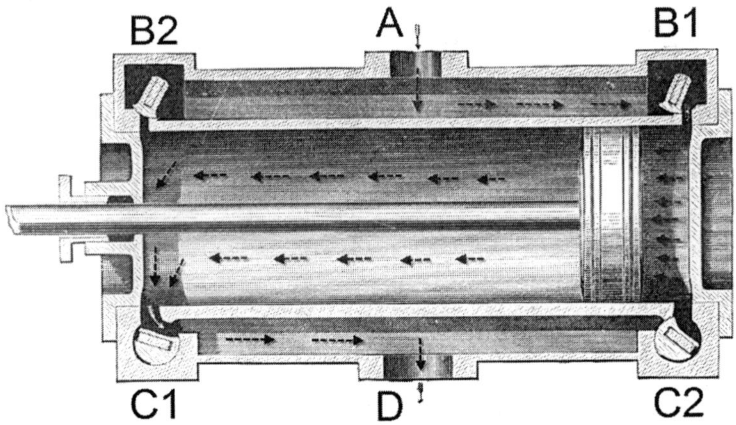

THE CORLISS ENGINE CYLINDER.

Figure: Detail of a Corliss-type valve gear and cylinder cross section showing the path of high-pressure steam (in red) and low-pressure steam (in blue). With each stroke, the four valves alternate opening and closing, driving the piston back and forth

As the area of a Corliss valve is small compared to the port area, the effects of gas flow generate relatively little torque on the valve axle compared to some other sorts of valve. These advantages have led to the Corliss form of valve being used in other roles, apart from steam engines with Corliss gear.

The Rolls-Royce Merlin aero-engine used a rectangular butterfly valve as a throttle. Gas-flow forces acting asymmetrically on this butterfly could lead to poor control of the power in some circumstances. Late models, from the 134, used a Corliss throttle valve instead to avoid this problem.

Barring and Barring Engines

A common feature of large Corliss engines is one or two sets of narrow gear teeth in the rim of the flywheel. These teeth allow the flywheel to be *barred*, that is, turned with the aid of a crowbar. This may be needed during engine maintenance, for example, to set the cutoff and admission valve timing, and it may be needed during engine starting.

The need for barring the engine during starting is most obvious on single-cylinder engines, where a careless engine operator might stop the engine with piston in dead centre. Once stopped in dead centre, the engine cannot be started under its own power, so it must be barred to a more favourable position for starting.

Large Corliss engines cannot be safely started cold, so it is common to admit low-pressure steam to both sides of the cylinder to warm up the metalwork. Turning the engine slowly during this process ensures that the entire engine is uniformly warmed, and it ensures that oil is uniformly distributed through the mechanism before applying power. Again, barring may be used to do this, although operators sometimes do this by careful manual manipulation of the valves.

For large engines, muscle powered barring is sufficiently difficult that *barring engines* are frequently installed. These are small engines with gear teeth cut to mate with the teeth on the flywheel. Generally, the drive gears of the barring engine are designed to automatically disengage if the engine begins running under its own power while the barring gears are engaged.

Company History

The Corliss Steam Engine Company was originally known as Fairbanks, Clark & Co. in the 1830s. In 1843 it was renamed Fairbanks, Bancroft & Co. when Edward Bancroft joined the company. In 1846 it was renamed Bancroft, Nightingale & Co. when George H. Corliss joined the company, and in 1847 it was renamed Corliss, Nightingale and Co. In 1848 the company moved to the Charles Street Railroad Crossing in Providence, Rhode Island.

In 1857 the company was renamed for the last time to Corliss Steam Engine Company. By 1864 Corliss bought out his partners and was the sole owner of the company. In 1900 the Corliss Steam Engine

Company was purchased by the International Power Company. In 1905 it was purchased by the American and British Manufacturing Company. In 1925 the company merged into Franklin Machine Company. By then Franklin Machine Company was already owned by the William A. Harris Steam Engine Company.

By 1859, all of the key features of what we now know as the Corliss engine were in place. Patents granted to Corliss and others incorporated rotary valves and crank shafts in-line with the cylinders. Competing inventors worked hard to invent alternatives to Corliss' mechanisms; they generally avoided Corlis's use of a wrist plate and adopted alternative releasing mechanisms for the steam valves, as in Jamieson's U.S. Patent 19,640, granted March 16, 1858.

Corliss' 1849 patent expired in 1870; the term of this patent had been extended by U.S. Patent reissue 200 on May 13, 1851 and U.S. Patent reissues 758 and 763 on July 12, 1859. After 1870, numerous other companies began to manufacture Corliss engines. Among them, the William A. Harris Steam Engine Company, the Worthington Pump and Machinery Company 2, and Allis-Chalmers 3. In general, these machines were referred to as Corliss engines regardless of who made them.

Centennial Engine

The Corliss Centennial Engine was an all-inclusive, specially built rotative beam engine that powered virtually all of the exhibits at the Centennial Exposition in Philadelphia in 1876 through shafts totaling over a mile in length. Switched on by President Ulysses Grant and Emperor Dom Pedro of Brazil, the engine was in public view for the duration of the fair.

The engine was configured as two cylinders side-by-side. Each cylinder was bored to 44 inches (1.1 m) with a stroke of 10 feet (3.0 m), making it the largest engine of the nineteenth century. The Centennial Engine was 45 feet (14 m) tall, had a flywheel 30 feet (9.1 m) in diametre, and produced 1,400 hp. After the fair it was disassembled and shipped back to Corliss's plant in Providence. Seven years later it was sold and powered a Chicago factory owned by George Pullman until 1910, when it was sold as scrap.

This engine became a cultural icon, so much so that to many modern historians the term Corliss Engine (or Corliss Steam Engine) refers to this specific engine and not to the broad class of engines fitted with Corliss valve gear.Tappet

A tappet is the noise made by a worn cam follower, however the term is widely used to represent the cam follower itself. In mechanical engineering it is a projection which imparts a linear motion to some other component within an assembly. Properly speaking, a tappet is the only part of a rocker arm that makes contact with an intake or exhaust valve stem above the cylinder head of an internal combustion engine. As the cam rotates, it creates both a sideways and a downward force on the tappet. Without a tappet (and with the cam acting directly on the valve), the sideways force would cause the valve stem to bend. With a tappet, the sideways force is transferred to the cylinder head so only the downward force acts on the valve stem.

Tappets/lifters are engineered to last the lifetime of the engine when the engine is properly maintained, including proper oil and filter changes. Though they are reliable, they have their drawbacks; valve clearances periodically require adjustment as the contact surfaces of both tappet and valve stem wear.

On most modern overhead cam engines, the camshaft is directly over the valves, as pictured. In some other overhead cam engines, a rocker arm pivots on a fixed shaft while one projection of the rocker-arm rides on a cam of the rotating camshaft. In both cases, this creates an oscillating linear motion, opening the valve. The closing of the valve is typically accomplished by a compression spring placed between the valve collet (or retainer) and the cylinder head above the combustion chamber. In overhead valve engines, the camshaft is located near the crankshaft and motion of the cam lobes is followed by cam followers (similar to tappets) and transferred up to the cylinder head and the rocker arm assembly by means of long pushrods.

Traditionally, the nominal distance (clearance) between the tappet surface and the valve's contact surface was maintained by means of an adjustment screw on the rocker arm. Today, this is typically accomplished by introducing shims into the space between the cam follower (tappet) and cam to create the necessary clearance, or by hydraulic adjusters. In the future, the use of tappets may stop entirely as engines move to variable valve timing systems.

Hydraulic Tappet

A hydraulic valve lifter, also known as a hydraulic tappet or a hydraulic lash adjuster, is a device for maintaining zero valve clearance in an internal combustion engine. The conventional means of adjusting valve actuation always requires a small clearance to be left between the valve and its rocker or cam follower to allow for thermal expansion and wear. The hydraulic lifter was designed to ensure that the valve train always operates with zero clearance, leading to quieter operation and eliminating the need for periodic adjustment of valve clearance.

The hydraulic lifter consists of a hollow expanding piston situated between the camshaft and valve. It is operated either by a rocker mechanism, or in the case of one or more overhead camshafts, directly by the camshaft. The lifter is filled with engine oil intermittently from an oil gallery via a small drilling. When the engine valve is closed, the lifter is free to fill with oil. When the valve is opening and the lifter is being operated by the camshaft, the oil feed is blocked and the lifter acts just as a solid one would, oil being incompressible.

History

The first firm to include hydraulic lifters in its design was Pierce-Arrow in the early 1930s. Hydraulic lifters were popular on automobiles designed in the 1980s, but most newer cars have reverted to bucket-and-shim mechanical lifters. Although these do not run as quietly and are not maintenance-free, they are cheaper and rarely need adjustment because the wear caused by operation is spread over a large area.

Disadvantages

There are a number of potential problems with hydraulic lifters. Frequently, the valvetrain will rattle loudly on startup due to oil draining from the lifters when the vehicle is parked. This is not considered significant provided the noise disappears within a couple of minutes, typically it usually only lasts a second or two. A rattle that does not go away can indicate a blocked oil feed or that one or more of the lifters has collapsed due to wear and is no longer opening its valve fully. The affected lifter should be replaced in the latter situation. In certain circumstances, a lifter can "pump up" and create negative valve clearance so that its valve cannot close. Typically this occurs in the uppermost speed range of an engine, and limits the engine's performance. Lifter pump-up is serious, as there may be interference of the valve with the piston or, burned valves may result. In all cases it is important to follow the manufacturer's recommendations for oil viscosity and quality.

Active Valve Control System

The active valve control system (AVCS) is an automobile variable valve timing technology used by Subaru. It varies the timing of the intake valves by using hydraulic oil pressure to rotate the camshaft in order to provide optimal air flow in and out of the engine. The system is closed loop using camshaft sensors, crankshaft sensors, air flow metre, throttle position as well as oxygen sensors and/or Air-Fuel ratio sensors in order to calculate engine load. The ECU is programmed to operate control valves that adjust the delivery of the hydraulic pressure in order to move the camshaft into the position that will provide the engine with the best performance while minding emissions standards.

AVCS is used on the Version 7 and up EJ207 engines, the EJ255 and EJ257, the second generation EZ30D (2005+ in the USA Market) found in the Legacy Outback, Legacy 3.0R and the B9 Tribeca. A Dual AVCS system utilises both the intake and exhaust camshafts on the EZ36 found in the 2008 Tribeca. Dual AVCS is also found on Japanese Spec engines, although a complete list of the engines with Dual AVCS has yet to be compiled.

i-AVLS

i-Active Valve Lift System (i-AVLS) is a version that supports variable valve lift. It includes 2 intake lift profiles depending on engine speeds. AVLS-equipped Subaru engines have 2 intake valves per cylinder, but AVLS only controls one of them, leaving the other in high-lift mode at all times.

At low engine speeds, each of the two intake valves for a given cylinder run on a different camshaft profile (the engines are 4-valve per cylinder type). One valve is operated by a low-lift, low-duration lobe, while the other is actuated by a high-lift, high-duration lobe. This increases air speed causing a smooth idle, while at the same time promoting a swirl factor, increasing efficiency.

At high speeds, the ECU (engine control unit) computer signals a solenoid to let oil pressure move a pin which locks the two lobes together. This means at high speed both intake valves are effectively driven by the high-speed camshaft lobe. This increases high-RPM power due to increased valve lift, duration and overall timing.

Cam locking is done by moving a pin with hydraulic pressure. As with other variable valve timing systems, it allows the engine to have a quiet low-speed operation with a smooth idle, while also using a camshaft lobe profile similar to “race cam” at high engine RPMs.

Applications

It was introduced with 2007 model year with 2.5-litre naturally aspirated Subaru engines. Widespread use of AVLS began with non-turbocharged Subaru 2.5 litre four-cylinder engines in Model year

2006. AVLS was first used in the 2005 six-cylinder 3.0 litre Subaru Outback. Turbocharged engines and Subaru's six-cylinder engines continue to use Subaru's AVCS (variable cam phasing) system instead of AVLS. Reports generally agree the new engine feels like a brand new engine with flatter torque curve compared to previous non-AVLS, non-AVCS 2.5 litre engines. Initial applications raised maximum power from 165 hp (123 kW) to 175 hp (130 kW), but more significantly raised the power at most RPM levels. The goal is better fuel economy, improved driveability and increased power.

Model Years

- 2005: 3.0 litre naturally aspirated H6 Subaru engines (used along with AVCS)
  - 250 hp (186 kW), 219 lb·ft (297 N·m) torque in 2005 3.0-litre Subaru Outback
- 2006: all 2.5 litre naturally aspirated Subaru engines
  - 173 hp (129 kW), 166 lb·ft (225 N·m) torque in 2006 Subaru Forester and Impreza 2.5i models
  - 175 hp (130 kW), 169 lb·ft (229 N·m) torque in 2007 Subaru Legacy model
- 2007–: all 2.5-litre naturally aspirated Subaru engines

Advanced VTEC

Advanced VTEC is a system designed to allow the variation of valve timing and lift. It is designed to improve engine Fuel efficiency without sacrificing performance. It is an evolution of a long line of VTEC systems from Honda.

Differences from Other VTECs

Honda's advanced VTEC technology departs greatly from its previous incarnations by no longer relying on switching between two sets of lobes on a given camshaft. It instead uses a single cam lobe per valve, and two rocker arms per valve whereby the second rocker arm has a movable pivot point, there by provide the varying cam lift. Advanced VTEC motors still use the now standard oil pressure controlled variable cam gear angle mechanism. With these two technologies combined Honda has developed an infinitely variable valve timing and lift system ("VVTL"). Previous versions of VTEC included only staged VVTL i.e. High-Low. With the introduction of i-VTEC the systems gained infinitely variable valve timing but still only staged lift i.e. High-Low. The "infinitely variable" portion of the

A-VTEC is what makes it stand out as a serious evolutionary step in the world of VTEC.

Explanation of Patent

Advanced VTEC has a standard camshaft and rocker arms, attached as they normally are with camshaft overhead, and rocker arms pushing down on the poppet valves. The camshaft is surrounded by a partially open drum which has secondary rocker arms attached to it via a pivoting point. These secondary rocker arms, which have a varying depth profile (similar to cams), are directly actuated by the camshaft, in a scissor-like manner. The primary rocker arms are actuated by the secondary (drum attached) rocker arms. The drum will only rotate to advance or retard the position of the secondary rocker arms, to take advantage of their varying profiles. Thus, through varying the position of the drum about its axis, each cam profile is changed to an optimal height for maximum engine performance without sacrificing fuel efficiency at lower speeds.

Camcon Binary Actuator

The Camcon Binary Actuator is a digital valve device used by oil and gas operators and sold by Camcon Technology.

MIVEC

MIVEC (Mitsubishi Innovative Valve timing Electronic Control system) is the brand name of a variable valve timing (VVT) engine technology developed by Mitsubishi Motors. MIVEC, as with other similar systems, varies the timing of the intake and exhaust camshafts which increases the power and torque output over a broad engine speed range while also being able to help spool a turbocharger more quickly.

MIVEC was first introduced in 1992 in their *4G92* powerplant, a 1,597 cc naturally aspirated DOHC 16 valve straight-4. At the time, the first generation of the system was named *Mitsubishi Innovative Valve timing and lift Electronic Control*. The first cars to use this were the Mitsubishi Mirage hatchback and the Mitsubishi Lancer sedan. While the conventional *4G92* engine provided 145 PS (107 kW; 143 hp) at 7000 rpm, the MIVEC-equipped engine could achieve 175 PS (129 kW; 173 hp) at 7500 rpm. Similar improvements were seen when the technology was applied to the 1994 Mitsubishi FTO, whose top-spec GPX variant had a *6A12* 1997 cc DOHC 24 valve V6 with peak power of 200 PS (147 kW; 197 hp) at 7500 rpm. The GR model, whose otherwise identical powerplant was not MIVEC-equipped, produced 170 PS (125 kW; 168 hp) at 7000 rpm by comparison.

Although initially designed to enhance performance, the system has subsequently been developed to improve economy and emissions, and has been introduced across Mitsubishi's range of vehicles, from the i kei car to the high-performance Lancer Evolution sedan.

Newest developments have led to MIVEC system being evolved into a continuous variable valve timing and also being the first VVT system to be used into a passenger car diesel engine.

Operation

Some types of variable valve control systems optimize power and torque by varying valve opening times and/or duration. Some of these valve control systems optimize performance at low and mid-range engine speeds. Others focus on enhancing only high-rpm power. MIVEC system provides both of these benefits by controlling valve timing and lift. The basic operation of the MIVEC system is altering the cam profiles and thus tailoring engine performance in response to driver input.

In essence, MIVEC serves the same function as "swapping cams", something that car racers might do when modifying older-design engines to produce more power. However, such swaps come with a compromise - generally yielding either greater low-end torque or more high-end horsepower, but not both. MIVEC achieves both goals. With MIVEC, the "cam swap" occurs automatically at a fixed engine speed. The cam switch operation is transparent to the driver, who is simply rewarded with a smooth flow of power.

Two distinct cam profiles are used to provide two engine modes: a low-speed mode, consisting of low-lift cam profiles; and a high-speed mode. The low-lift cams and rocker arms - which drive separate intake valves - are positioned on either side of a centrally located high-lift cam. Each of the intake valves is operated by a low-lift cam and rocker arm, while placing a T-lever between them allows the valves to follow the action of the high-lift cam.

At low speeds, The T-lever's wing section floats freely, enabling the low-lift cams to operate the valves. The intake rocker arms contain internal pistons, which are retained by springs in a lowered position while the engine speed is below the MIVEC switchover point, to avoid contacting the high-lift T-shaped levers. At high speeds, hydraulic pressure elevates the hydraulic pistons, causing the T-lever to push against the rocker arm, which in turn makes the high-lift cam operate the valves.

In summary, MIVEC switches to the higher cam profile as engine speed increases, and drops back to the lower cam profile as engine speed decreases. The reduced valve overlap in low-speed mode provides stable idling, while accelerated timing of the intake valve's closing reduces backflow to improve volumetric efficiency, which helps increase engine output as well as reduce lift friction. High-speed mode takes advantage of the pulsating intake effect created by the mode's high lift and retarded timing of intake valve closure. The resulting reduced pumping loss of the larger valve overlap yields higher power output and a reduction in friction. The low- and high-speed modes overlap for a brief period, boosting torque.

From the 4B1 engine family onward, MIVEC has evolved into a continuous variable valve timing (CVVT) system (dual VVT on intake and exhaust valves). Many older implementations only vary the valve timing (the amount of time per engine revolution that the intake port is open) and not the lift. Timing is continuously independently controlled to provide four optimized engine-operating modes:

- Under most conditions, to ensure highest fuel efficiency, valve overlap is increased to reduce pumping losses. The exhaust valve opening timing is retarded for higher expansion ratio, enhancing fuel economy.
- When maximum power is demanded (high engine speed and load), intake valve closing timing is retarded to synchronize the intake air pulsations for larger air volume.
- Under low-speed, high load, MIVEC ensures optimal torque delivery with the intake valve closing timing advanced to ensure sufficient air volume. At the same time, the exhaust valve opening timing is retarded to provide a higher expansion ratio and improved efficiency.
- At idle, valve overlap is eliminated to stabilise combustion.

Mitsubishi's *4N1* engine family is the world's first to feature a variable valve timing system applied to passenger car diesel engines.

MIVEC-MD

In the early years of developing its MIVEC technology, Mitsubishi also introduced a variant dubbed MIVEC-MD (Modulated Displacement), a form of variable displacement. Under a light throttle load, the intake and exhaust valves in two of the cylinders would remain closed, and the reduced pumping losses gave a claimed 10–20 percent improvement in fuel economy. Modulated Displacement was dropped around 1996.

Multiair

Multiair (or Uniair) is an electro-hydraulic variable valve actuation technology controlling air intake (without a throttle valve) in petrol or diesel engines. It is compatible with both naturally aspirated and forced-induction engines.

Multiair technology was patented by Fiat in 2002; it was launched at the 2009 Geneva Motor Show to be available in the Alfa Romeo MiTo. The new engine is produced in Termoli, Italy at the Fiat Powertrain Technologies factory. The 1.4 L Multiair engine won the "Best New Engine of 2010" award. For the Fiat 500 North American version the 1.4 L engine is produced also at the Global Engine Manufacturing Alliance in Dundee, Michigan, United States. The Fiat TwinAir 875 cc two-cylinder engine is also using Multiair technology, it was named as Best New Engine of 2011. Technology

The basic idea, controlling intake valves, works upon the following principle: a valve tappet (cam follower), moved by a mechanical intake cam, is connected to the intake valve through a hydraulic chamber, controlled by a normally open on/off solenoid valve. The system allows optimum intake valve opening schedules.

An example of the gains offered by the MultiAir technology can be seen in the current Alfa Romeo MiTo: the 1.4 JTB engine, without MultiAir, can produce 155 PS (114 kW; 153 hp) and 230 N·m (170 lb·ft) whilst using 6.5 litres per 100 kilometres (43 mpg_{-imp}; 36 mpg_{-US}). With MultiAir, the same engine develops 170 PS (125 kW; 168 hp) and 250 N·m (180 lb·ft), yet only consumes 6.0 L/100 km (47 mpg_{-imp}; 39 mpg_{-US}). MultiAir engines will increase power (up to 10%) and torque (up to 15%), as well as considerably reduce consumption levels (up to 10%) and emissions of CO_2 (up to 10%), particulates (up to 40%) and NOx (up to 60%) when compared to a traditional petrol engine. The system also provides smoother cold weather operation, more even torque delivery and no engine shake at shut-off.

The Multiair technology took over a decade to develop. The vice president of Fiat Powertrain Research & Development Rinaldo Rinolfi led the team who developed the technology. Development costs were over $100 million. There was also delay in development, in the time (2000-2005) when Fiat was in partnership with General Motors.

Other Systems

More advanced, fully camless valvetrain systems are under development, but are not yet production-ready. The Valvetronic system used by BMW allows the valve timing and lift to be varied but not the cam profile. The ability to vary the latter is characteristic of camless and the Multiair systems.

N-VCT

Nissan Variable Cam Timing (commonly known as N-VCT, VCT, NVCS or CVTC) is an automobile variable valve timing technology developed by Nissan. N-VCT was first introduced in 1987 on the VG30DE and VG20DET engine.

N-VCT varies valve timing by rotating the affected camshaft(s) relative to the sprocket; valve lift and duration are *not* altered. This rotation is achieved when an electric solenoid, controlled by the car's ECU, allows pressurized engine oil to flow into and through the cam and into a slave mechanism, axially advancing cam timing relative to the sprocket. Some Nissan engines only have N-VCT on the inlet cam (such as the GA16DE or RB25DE/DET) while others have it on both the inlet and exhaust cams.

Engines That Have N-VCT

- RB25DE (1993+)
- RB25DET(1993+)
- GA16DE
- SR20DE
- SR20DET
- VQ35DE
- VQ40DE
- VE30DE
- VG20DET
- VG30DE
- VG30DETT
- VH45DE
- VK45DE
- QG18DE
- QG16DE

Nissan VVL Engine

Nissan Ecology Oriented Variable Valve Lift and Timing (commonly known as VVL) is an automobile variable valve timing technology developed by Nissan. VVL varies the timing, duration, and lift of valves by using hydraulic pressure switch between two different sets of camshaft lobes. It functions similarly to Honda's VTEC system.

The SR20VE is the most common engine with NEO VVL. There have been two main versions of this engine. The first version made 187 hp (139 kW) and 145 lb·ft (197 N·m) torque. This engine was used by Nissan from 1997 to 2001. It is found in the Nissan Primera, Nissan Bluebird, and the Nissan Wingroad.

The second variant of the SR20VE is found only in the 2001 and later P12 Nissan Primera. This version of the SR20VE makes 204 hp (152 kW), and 152 lb·ft (206 N·m) torque. This engine is commonly known as the SR20VE '20V'. Although, in automotive terms, '20V' would normally be interpreted as having twenty valves, this is incorrect. The name '20V' is the name of the trim level of the Nissan Primera that it is found in. It is also a shortened version of the name SR20VE. This engine has 16 valves like the rest of the SR20 engines. This newer '20V' engine, is the only SR20 engine to get a restyled valve cover. It also has an upgraded intake manifold, which has longer runners and a larger 70 mm (2.8 in) throttle body (earlier SR20VE has 60 mm).

Another version of the VVL SR engines, is the 1.6 L SR16VE. The engine block for the SR16VE is the same as the SR20VE; it also has the same cylinder bore. The crankshaft has a shorter stroke, which lowers the displacement, but allows the engine to safely rev to higher RPM. Although this engine has 1.6 L of displacement, it has more aggressive camshaft specifications. It manages to make 173 hp (129 kW). The camshafts from this engine are considered to be an upgrade for SR20VE owners.

From 1997 to 1998, Nissan produced 500 limited-edition SR16VE N1 engines. These engines made 197 hp (147 kW). They had further upgraded camshafts, upgraded intake manifold using eight injectors and a larger 70 mm (2.8 in) throttle body. These engines were found in the limited-edition Nissan Pulsar VZ-R N1. They were only sold in Japan.

The most powerful VVL engine so far is the SR20VET. The SR20VET is a turbocharged '20V' SR20VE. It uses a Garrett GT2560R turbocharger, and makes 280 PS (206 kW; 276 hp). Nissan's technical information about this engine states that it is 9:1 compression ratio, but it really adds up to 8.8:1. Compared to the SR20DET (used in the Nissan Silvia, and Bluebird), the SR20VET (aside from having VVL technology) has improved airflow in the cylinder head, higher compression, and also improved coolant passages.

In 1998 Nissan's RB engine was also improved using the VVL NEO Head Technology, which provided the motor better fuel economy and lower exhaust emissions to keep up with emission standards.

One difference from Honda's VTEC system is that NEO VVL engages the change of intake and exhaust cams independently for a flatter, more consistent power band. On the SR20VE, the intake camshaft is switched at 5000 rpm, and the exhaust at 6500 rpm. However this trait was not included on the newer '20V' version, as both camshafts engage at the same RPM.

Valvetronic

VANOS (abbr. from German *variable Nockenwellensteuerung*) is an automobile variable valve timing technology developed by BMW in close collaboration with Continental Teves. VANOS varies the timing of the valves by moving the position of the camshafts in relation to the drive gear. This movement varies from 6 degrees of advanced to 6 degrees of retarded camshaft timing.

Single VANOS

VANOS is a combined hydraulic and mechanical camshaft control device managed by the car's DME engine management system. The VANOS system is based on a discrete adjustment mechanism that can modify the position of the intake camshaft versus the crankshaft. Double-VANOS adds continuous adjustability to the intake and exhaust camshafts.

VANOS operates on the intake camshaft in accordance with engine speed and accelerator pedal position. At the lower end of the engine-speed scale, the intake valves are opened later, which improves idling quality and smoothness. At moderate engine speeds, the intake valves open much earlier, which boosts torque and permits exhaust gas re-circulation inside the combustion chambers, reducing fuel consumption and exhaust emissions. Finally, at high engine speeds, intake valve opening is once again delayed, so that full power can be developed. VANOS significantly enhances emission management, increases output and torque, and offers better idling quality and fuel economy.

VANOS was first introduced in 1992 on the BMW M50 engine used in the 5 Series. VANOS was earlier available from Alfa Romeo. The 2 liter "Nord" engine, used in all Spiders, most GTV's and various coupes and sedans had VANOS. The early systems(early 1980's) were strictly mechanical (centrifugal), and the later systems were electric activated oil pressure operated systems. This was probably the first VANOS available on a production car.

Double VANOS

The first double VANOS system appeared on the S50B32 engine in 1996. Later, BMW added "double" VANOS to its M52TU series of inline 6-cylinder engines, which changed the mechanism from fixed position operation to continuously variable, and added the same functionality to the exhaust camshaft, on a number of its cars. Double-VANOS (double-variable camshaft control) significantly improves torque and emissions since valve timing on both the intake and exhaust camshafts is adjusted to the power required from the engine as a function of gas pedal position and engine speed. On all BMW engines that use single VANOS, except S50B30, the timing of the intake cam is only changed at two distinct rpm points, while on the double-VANOS system, the timing of the intake and exhaust cams is continuously variable through a range of ~40 crankshaft degrees for the intake, and 25 degrees for the exhaust. The advantage of double-VANOS is that the system controls the flow of hot exhaust gases into the intake manifold individually for all operating conditions. This is referred to as "internal" exhaust gas re-circulation, allowing very fine dosage of the amount of exhaust gas recycled.

VANOS Performance

Single and Double VANOS systems due to age and mileage often experience performance issues as well as complete failure of the BMW OEM Buna-N VANOS piston seals (which are often successfully repaired by the use of aftermarket Viton seals to replace the OEM Buna-N seals). Another common effect is a rattling sound coming from the upper valve train assembly. The original seal material was found to be inferior for the application and tends to break down within 75k - 100k miles. VANOS Seal & Rattle kits are available to address these shortcomings from aftermarket suppliers (but not from BMW — BMW's solution when VANOS units fail is to remove the entire VANOS unit and replace it with a new one). VANOS repair may require special tools (cooling fan clutch wrenches, etc.) that can be rented or purchased from a variety of vendors.

List of VANOS engines

Name	VANOS Type	Infinite Adjustments	Solenoids
M50TU	Single	No	1 solenoid I/O
M52	Single	No	1 solenoid I/O
M52TU	Double	Yes	2 solenoid pwm
M54	Double	Yes	2 solenoid pwm
N52	Double	Yes	2 solenoid pwm
S50	Single	Yes	1992-1994 not in u.s - 2 solenoid pwm
S50	Double	Yes	1995-1998 not in u.s - 4 solenoid pwm
S52	Single	No	1996-1998 u.s - 1 solenoid I/O
S54	Double	Yes	- 4 solenoid pwm
M62TU	Single	Yes	- 2 solenoid pwm
N62	Double	Yes	- 4 solenoid pwm
S62	Double	Yes	- 8 solenoid pwm + 1 pwm solenoid pressure regulator
S85	Double	Yes	- 4 solenoid pwm + 1 pwm pressure regulator
N73	Double	Yes	- 4 solenoid pwm

Note: The engines starting with an "N" also have valvetronic variable intake valve opening.

Variable Cam Timing

Variable Camshaft Timing (VCT) is an automobile variable valve timing technology developed by Ford. It uses electronically controlled hydraulic valves that direct high pressure engine oil into the camshaft phaser cavity. These oil control solenoids are bolted into the cylinder heads towards the front of the engine near the camshaft phasers. The powertrain control module (PCM) transmits a signal to the solenoids to move a valve spool that regulates the flow of oil to the phaser cavity. The phaser cavity changes the valve timing by rotating the camshaft slightly from its initial orientation, which results in the camshaft timing being advanced or retarded. The PCM adjusts the camshaft timing depending on factors such as engine load and RPM. This allows for more optimum engine performance, reduced emissions, and increased fuel efficiency compared to engines with fixed camshafts.

For twin cam or DOHC engines, VCT was used on either the intake or exhaust camshaft. (Engines that have VCT on both camshafts are now designated as Ti-VCT.) The use of variable camshaft timing on the exhaust camshaft is for improved emissions, and vehicles with VCT on the exhaust camshaft do not require exhaust gas recirculation (EGR) as retarding the exhaust cam timing achieves the same result. VCT on the intake camshaft is used primarily for increasing engine power and torque as the PCM is able to optimize the opening of the intake valves to match the engine conditions.

VCT is used in Ford's Triton 5.4L 3-valve V8 engine, the Australian Barra 182 and 240 Inline-6s, and Ford's 4.6L 3-valve V8 engine used in the 2006-2010 Ford Explorer and 2005-2010 Ford Mustang GT.

The 2.0L Zetec Inline-4 used in the 1998-2003 Ford Escort ZX2, Ford Contour, and 99-02 Mercury Cougar used VCT on the exhaust camshaft. The 2002-2004 SVT Focus (ST170 in Europe) also featured VCT, but on the intake camshaft of its modified version of the 2.0L Zetec engine. In addition, the 1.7L Zetec-S engine found in the European Ford Puma was equipped with variable camshaft timing. The 6.2L V8 introduced in the 2010 SVT Raptor also uses VCT. That motor has a single cam per bank, so it is dual-equal variable cam timing.

Ti-VCT

Twin Independent Variable Camshaft Timing (Ti-VCT) is the name given by Ford to engines with the ability to vary the camshaft timing of both the intake and exhaust camshafts independently. Despite the name difference, its operation is the same as standard VCT. Unlike the original versions of VCT, which only operated on a single camshaft, Ti-VCT allows the PCM to advance or retard the timing of both the intake and exhaust camshafts independently. This allows for improved power and torque, particularly at lower engine RPM, as well as improved fuel economy and reduced emissions.

Many new Ford engines feature Ti-VCT, including the 2011 Mustang, 2011 Edge and Edge Sport, 2011 Lincoln MKX, 2011 Fiesta, 2011 Explorer, and 2012 Focus.

Variable Valve Control

VVC (Variable Valve Control) is an automobile variable valve timing technology developed by Rover and applied to some high performance variants of the company's K Series 1800cc engine.

In order to improve the optimisation of the valve timing for differing engine speeds and loads, the system is able to vary the timing and duration of the inlet valve opening. It achieves this by using a complex and finely machined mechanism to drive the inlet camshafts. This mechanism can accelerate and decelerate the rotational speed of the camshaft during different parts of its cycle. e.g. to produce longer opening duration, it slows the rotation during the valve open part of the cycle and speeds it up during the valve closed period.

The system has the advantage that it is continuously variable rather than switching in at a set speed. Its disadvantage lies in its complexity. Other systems achieve similar results with less expense and better reliability. The VVC engine is also fitted into these non MG/Rover cars: Ariel Atom, Lotus Elise, Caterham 7, Caterham 21, GTM Libra

Variable Valve Event and Lift

Nissan Variable Valve Event and Lift (commonly abbreviated as "VVEL") is an automobile variable valve timing technology developed by Nissan.

Nissan VVEL was first introduced to the US market in late-2007 on the 2008 Infiniti G37 Coupe sporting the new "VVEL" VQ37VHR Nissan VQ Engine (VQ37VHR motor specs: 11.0:1 CR, 95.5mm bore, 86mm stroke, 7500rpm redline). VVEL variable valve timing is now also used in the VK50VE V8 engine from the Infiniti FX50.

A rocker arm and two types of links open the intake-valves by transferring the rotational movement of a drive shaft with an eccentric cam to the output cam. The movement of the output cam is varied by rotating the control shaft with a DC stepper motor and changing the fulcrums of the links. This makes the continuous adjustment of the amount of the valve's lift (e.g., the amount of intake opening), during the intake valve event in the four-stroke cycle, possible. C-VTC and VVEL together control the valve phases and its valve events, allowing free-control of the valve timing and lift. This results in more efficient airflow to the cylinder, significantly improved responsiveness, optimizing the balance between power and environmental performance.

It performs similarly to BMW's Valvetronic system but with desmodromic control of the output cam, allowing VVEL to operate at higher engine speeds (RPM). Other similar systems are offered Toyota (Valvematic).

VarioCam

VarioCam is an automobile variable valve timing technology developed by Porsche. VarioCam varies the timing of the intake valves by adjusting the tension on the timing chain connecting the intake and exhaust camshafts. VarioCam was first used on the 1992 3.0 L engine in the Porsche 968.

Porsche's more recent VarioCam Plus combines variable valve timing with two-stage lift on the intake side. The two-stage valve-lift function is performed by electro-hydraulically operated switchable tappets. Each of these 12 tappets consists of concentric lifters which can be locked together by means of a pin. The inner lifter is actuated by a small cam lobe, while the outer ring element is moved by a pair of larger-profile lobes. The timing of each valve is seamlessly adjusted by means of an electro-hydraulically operated rotary vane adjuster at the head of each intake camshaft.

Valve timing and the valve profile are continuously altered according to conditions and engine load. For improved responsiveness on cold starts, VarioCam Plus raises the amount of lift and retards valve timing. At medium revs with minimal loads, the valve lift is lowered and timing advanced to help minimise fuel consumption and emissions. For maximum power and torque, the lift is raised and the timing is advanced. This system debuted on the 1999 Porsche 996 Turbo.

A system similar to the VarioCam Plus system was developed by Porsche for Subaru for the redesigned EZ30R H6 engine which debuted in 2003 in the Legacy and Outback.

VTEC

This article is about the valvetrain system. VTEC (Variable Valve Timing Engine Control) is a valvetrain system developed by Honda to improve the volumetric efficiency of a four-stroke internal combustion engine. The VTEC system uses two camshaft profiles and electronically selects between the profiles. It was invented by Honda engineer Ikuo Kajitani, and was the first system of its kind. Different types of variable valve timing and lift control systems have also been produced by other manufacturers (MIVEC from Mitsubishi, AVCS from Subaru, VVT-i/VVTL-i from Toyota, VANOS from BMW, VarioCam Plus from Porsche, NeoVVL from Nissan, etc.).

Context, and description

VTEC was initially designed to increase the power output of an engine to 100 HP/litre or more while maintaining practicality for use in mass production vehicles. Some later variations of the system were designed solely to provide improvements in fuel efficiency.

Japan levies a tax based on engine displacement, and Japanese auto manufacturers have correspondingly focused their R&D efforts toward improving the performance of smaller engine designs through means other than displacement increases. One method for increasing performance into a static displacement includes forced induction, as with models such as the Toyota Supra and Nissan 300ZX which used turbocharger applications and the Toyota MR2 which used a supercharger for some model years. Another approach is an unconventional engine design, such as the rotary engine used in the

Mazda RX-7 and RX-8. A third option is to change the cam timing profile, of which Honda VTEC was the first successful commercial design for altering the profile in real-time.

The VTEC system provides the engine with multiple camshaft profiles optimized for both low and high RPM operations. In basic form, the single cam profile of a conventional engine is replaced with two profiles: one optimized for low-RPM stability and fuel efficiency, and the other designed to maximise high-RPM power output. The switching operation between the two cam lobes is controlled by the ECU which takes account of engine oil pressure, engine temperature, vehicle speed, engine speed and throttle position. Using these inputs, the ECU is programmed to switch from the low lift to the high lift cam lobes when the conditions mean that engine output will be improved. At the switch point a solenoid is actuated which allows oil pressure from a spool valve to operate a locking pin which binds the high RPM cam follower to the low RPM ones. From this point on, the poppet valve opens and closes according to the high-lift profile, which opens the valve further and for a longer time. The switch-over point is variable, between a minimum and maximum point, and is determined by engine load. The switch-down back from high to low RPM cams is set to occur at a lower engine speed than the switch-up (representing a hysteresis cycle) to avoid a situation in which the engine is asked to operate continuously at or around the switch-over point.

The older approach to timing adjustments is to produce a camshaft with a valve timing profile that is better suited to high-RPM operation. The improvements in high-RPM performance occur in trade for a power and efficiency loss at lower RPM ranges, which is where most street-driven automobiles operate a majority of the time. Correspondingly, VTEC attempts to combine high-RPM performance with low-RPM stability.

History

VTEC, the original Honda variable valve control system, originated from REV (Revolution-modulated valve control) introduced on the CBR400 in 1983 known as HYPER VTEC. In the regular four-stroke automobile engine, the intake and exhaust valves are actuated by lobes on a camshaft. The shape of the lobes determines the timing, lift and duration of each valve. Timing refers to an angle measurement of when a valve is opened or closed with respect to the piston position (BTDC or ATDC). Lift refers to how much the valve is opened. Duration refers to how long the valve is kept open. Due to the behaviour of the working fluid (air and fuel mixture) before and after combustion, which have physical limitations on their flow, as well as their interaction with the ignition spark, the optimal valve timing, lift and duration settings under low RPM engine operations are very different from those under high RPM. Optimal low RPM valve timing, lift and duration settings would result in insufficient filling of the cylinder with fuel and air at high RPM, thus greatly limiting engine power output. Conversely, optimal high RPM valve timing, lift and duration settings would result in very rough low RPM operation and difficult idling. The ideal engine would have fully variable valve timing, lift and duration, in which the valves would always open at exactly the right point, lift high enough and stay open just the right amount of time for the engine speed in use.

DOHC VTEC

Introduced as a DOHC system in Japan in the 1989 Honda Integra XSi and the Honda CR-X SiR, which both used the 160 bhp (120 kW) B16A engine. The same year, Europe saw the arrival of VTEC in the Honda CRX 1.6i-VT, using a 150 bhp variant (B16A1). The US market saw the first VTEC system with the introduction of the 1991 Honda NSX, which used a 3-litre DOHC VTEC V6 with 280 bhp (210 kW). DOHC VTEC engines soon appeared in other vehicles, such as the 1992 Acura Integra GS-R (B17A1 1.7-litre engine), and later in the 1992 Honda Prelude VTEC (H22A 2.2-litre engine with 195 hp) and Honda Del Sol VTEC (B16A3 1.6-litre engine). The Integra Type R (1997–2001) available in the Japanese market produces 200 bhp (149 kW; 203 PS) using a B18C5 1.8-litre engine. Honda has also continued to develop other varieties and today offers several varieties of VTEC, such as i-VTEC and i-VTEC Hybrid.

SOHC VTEC

As popularity and marketing value of the VTEC system grew, Honda applied the system to SOHC (Single Over Head Cam) engines, which share a common camshaft for both intake and exhaust valves. The trade-off was that Honda's SOHC engines benefitted from the VTEC mechanism only on the intake valves. This is because VTEC requires a third centre rocker arm and cam lobe (for each intake and exhaust side), and, in the SOHC engine, the spark plugs are situated between the two exhaust rocker arms, leaving no room for the VTEC rocker arm. Additionally, the centre lobe on the camshaft cannot be utilised by both the intake and the exhaust, limiting the VTEC feature to one side.

However, beginning with the J37A4 3.7L SOHC V6 engine introduced on all 2009 Acura TL SH-AWD models, SOHC VTEC was incorporated for use with intake and exhaust valves. The intake and exhaust rocker shafts contain primary and secondary intake and exhaust rocker arms, respectively. The primary rocker arm contains the VTEC switching piston, while the secondary rocker arm contains the return spring. The term "primary" does not refer to which rocker arm forces the valve down during low-RPM engine operation. Rather, it refers to the rocker arm which contains the VTEC switching piston and receives oil from the rocker shaft.

The primary exhaust rocker arm contacts a low-profile camshaft lobe during low-RPM engine operation. Once VTEC engagement occurs, the oil pressure flowing from the exhaust rocker shaft into the primary exhaust rocker arm forces the VTEC switching piston into the secondary exhaust rocker arm, thus locking both exhaust rocker arms together. The high-profile camshaft lobe which normally contacts the secondary exhaust rocker arm alone during low-RPM engine operation is able to move both exhaust rocker arms together which are locked as a unit. The same occurs for the intake rocker shaft, except that the high-profile camshaft lobe operates the primary rocker arm.

The difficulty of incorporating VTEC for both the intake and exhaust valves in a SOHC engine has been removed on the J37A4 by a novel design of the intake rocker arm. Each exhaust valve on the J37A4 corresponds to one primary and one secondary exhaust rocker arm. Therefore, there are a total of twelve primary exhaust rocker arms and twelve secondary exhaust rocker arms. However, each secondary intake rocker arm is shaped similar to a "Y" which allows it to contact two intake valves at once. One primary intake rocker arm corresponds to each secondary intake rocker arm. As a result of this design, there are only six primary intake rocker arms and six secondary intake rocker arms.

VTEC-E

VTEC-E is a variation of SOHC VTEC which was used to increase efficiency at low RPM. At low RPM, one of the two intake valves is only allowed to open a very small amount, increasing the fuel/air atomization in the cylinder and thus allowing a leaner mixture to be used. As the engine's speed increases, both valves are needed to supply sufficient mixture. A sliding pin, which is pressured by oil, as in the regular VTEC, is used to connect both valves together and allows the full opening of the second valve. The engine runs at normal performance using the second cam position that would typically be tuned for high-RPM performance in other two-stage VTEC designs.

3-Stage VTEC

3-Stage VTEC is a version that employs 3 different cam profiles to control intake valve timing and lift. Due to this version of VTEC being designed around a SOHC valve head, space was limited and so VTEC can only modify the opening and closing of the intake valves. The low-end fuel economy improvements of VTEC-E and the performance of conventional VTEC are combined in this application. From idle to 2500-3000 RPM, depending on load conditions, one intake valve fully opens while the other opens just slightly, enough to prevent pooling of fuel behind the valve, also called 12-valve mode.

This 12 Valve mode results in swirl of the intake charge which increases combustion efficiency resulting in improved low end torque and better fuel economy. At 3000-5400 RPM, depending on load, one of the VTEC solenoids engages which causes the 2nd valve to lock onto the first valve's camshaft lobe.

Also called 4-valve mode, this method resembles a normal engine operating mode and improves the mid-range power curve. At 5500-7000 RPM, the second VTEC solenoid engages (both solenoids now engaged) so that both intake valves are using a middle, third camshaft lobe. The third lobe is tuned for high-performance and provides peak power at the top end of the RPM range.

i-VTEC

Honda i-VTEC (intelligent-VTEC) has VTC continuously variable timing of camshaft phasing on the intake camshaft of DOHC VTEC engines. The technology first appeared on Honda's K-series four-cylinder engine family in 2001 (2002 in the U.S.). In the United States, the technology debuted on the 2002 Honda CR-V.

VTC controls of valve lift and valve duration are still limited to distinct low- and high-RPM profiles, but the intake camshaft is now capable of advancing between 25 and 50 degrees, depending upon engine configuration. Phasing is implemented by a computer controlled, oil driven adjustable cam gear. Both engine load and RPM affect VTC. The intake phase varies from fully retarded at idle to somewhat advanced at full throttle and low RPM. The effect is further optimization of torque output, especially at low and midrange RPM. There are two types of i-VTEC K series engines which are explained in the next paragraph.

K-series

The K-Series motors have two different types of i-VTEC systems implemented. The first is for the performance motors like in the RSX Type S or the Civic Si and the other is for economy motors found in the CR-V or Accord. The performance i-VTEC system is basically the same as the DOHC VTEC system of the B16A's; both intake and exhaust have 3 cam lobes per cylinder. However the valvetrain has the added benefit of roller rockers and continuously variable intake cam timing. Performance i-VTEC is a combination of conventional DOHC VTEC with VTC. The economy i-VTEC is more like the SOHC VTEC-E in that the intake cam has only two lobes, one very small and one larger, as well as no VTEC on the exhaust cam. The two types of motor are easily distinguishable by the factory rated power output: the performance motors make around 200 hp (150 kW) or more in stock form and the economy motors do not make much more than 160 hp (120 kW) from the factory.

R-series

i-VTEC with Variable Cylinder Management (VCM)

In 2003, Honda introduced an i-VTEC V6 (an update of the J-series) that includes Honda's cylinder deactivation technology which closes the valves on one bank of (3) cylinders during light load and low speed (below 80 km/h (50 mph)) operation. According to Honda, "VCM technology works on the principle that a vehicle only requires a fraction of its power output at cruising speeds. The system electronically deactivates cylinders to reduce fuel consumption. The engine is able to run on 3, 4, or all 6 cylinders based on the power requirement. Essentially getting the best of both worlds. V6 power when accelerating or climbing, as well as the efficiency of a smaller engine when cruising."

i-VTEC VCM was also used in 1.3L 4-cylinder engines used in Honda Civic Hybrid.

i-VTEC i

It is a version of i-VTEC with direct injection.

It was first used in 2003 Honda Stream.

AVTEC

The AVTEC (Advanced VTEC) engine was first announced in 2006. It combines continuously variable valve lift and timing control with continuously variable phase control. Honda originally planned to produce vehicles with AVTEC engines within next 3 years. Although it was speculated that it would first be used in 2008 Honda Accord, the vehicle instead utilises the existing i-VTEC system.

A related US patent (6,968,819) was filed in 2005-01-05.

VTEC in motorcycles

Apart from the Japanese market-only Honda CB400SF Super Four HYPER VTEC, introduced in 1999, the first worldwide implementation of VTEC technology in a motorcycle occurred with the introduction of Honda's VFR800 sportbike in 2002. Similar to the SOHC VTEC-E style, one intake valve remains closed until a threshold of 7000 rpm is reached, then the second valve is opened by an oil-pressure actuated pin. The dwell of the valves remains unchanged, as in the automobile VTEC-E, and little extra power is produced but with a smoothing-out of the torque curve. Critics maintain that VTEC adds little to the VFR experience while increasing the engine's complexity. Honda seemed to agree as their VFR1200, a model announced in October 2009, came to replace the VFR800; which abandons the V-TEC concept in favour of a large capacity narrow-vee "unicam" (i.e. sohc) motor.

VVT-i

VVT-i, or Variable Valve Timing with intelligence, is an automobile variable valve timing technology developed by Toyota, similar in performance to the BMW's VANOS. The Toyota VVT-i system replaces the Toyota VVT offered starting in 24 December 1991 on the 5-valve per cylinder *4A-GE* engine. The VVT system is a 2-stage hydraulically controlled cam phasing system. The Toyota motors CEO has been reported to have said, "VVT is the heart of every modern Toyota!" VVT-i, introduced in 1996, varies the timing of the intake valves by adjusting the relationship between the camshaft drive (belt, scissor-gear or chain) and intake camshaft. Engine oil pressure is applied to an actuator to adjust the camshaft position. Adjustments in the overlap time between the exhaust valve closing and intake valve opening result in improved engine efficiency. Variants of the system, including VVTL-i, Dual VVT-i, Triple VVT-iE, and Valvematic, have followed.

VVTL-i

VVTL-i (Variable Valve Timing and Lift intelligent system) is an enhanced version of VVT-i that can alter valve *lift* (and *duration*) as well as valve timing. In the case of the 16 valve *2ZZ-GE*, the engine head resembles a typical DOHC design, featuring separate cams for intake and exhaust and featuring two intake and two exhaust valves (four total) per cylinder. Unlike a conventional design, each camshaft has two lobes per cylinder, one optimized for lower rpm operation and one optimized for high rpm operation, with higher lift and longer duration. Each valve pair is controlled by one rocker arm, which is operated by the camshaft. Each rocker arm has a slipper follower mounted to the rocker arm with a spring, allowing the slipper follower to freely move up and down with the high lobe without affecting the rocker arm. When the engine is operating below 6000-7000 rpm (dependent on year, car, and ECU installed), the lower lobe is operating the rocker arm and thus the valves, and the slipper-follower is freewheeling next to the rocker arm. When the engine is operating above the lift engagement point, the ECU activates an oil pressure switch which pushes a sliding pin under the slipper follower on each rocker arm. The rocker arm is now locked into slipper-follower's movements and thus follows the movement of the high rpm cam lobe, and will operate with the high rpm cam profile until the pin is disengaged by the ECU. The lift system is similar in principle to Honda VTEC operation.

Dual VVT-i

The Dual VVT-i system adjusts timing on both intake and exhaust camshafts. It was first introduced in 1998 on the RS200 Altezza's *3S-GE* engine.

Dual VVT-i is also found in Toyota's new generation V6 engine, the 3.5-litre *2GR-FE* first appearing on the 2005 Avalon. This engine can now be found on numerous Toyota and Lexus models. By adjusting the valve timing, engine start and stop occurs almost unnoticeably at minimum compression. Fast heating of the catalytic converter to its light-off temperature is possible, thereby reducing hydrocarbon emissions considerably. Most Toyota engines including the UR engines (V8), GR engines (V6), AR engines (Large I4), and ZR engines (Small I4) now use this technology.

VVT-iE

VVT-iE (Variable Valve Timing - intelligent by Electric motor) is a version of Dual VVT-i that uses an electrically operated actuator to adjust and maintain intake camshaft timing. The exhaust camshaft timing is still controlled using a hydraulic actuator. This form of variable valve timing technology was developed initially for Lexus vehicles. This system was first introduced on the 2007MY Lexus LS 460 as 1UR engine.

Bibliography

Arlosoroff, S.: *Community Water Supply: the Handpump Option*, World Bank, Washington, DC, 1987.

Ashby, Michael; Hugh Shercliff and David Cebon: *Materials: Engineering, Science, Processing and Design*, Butterworth-Heinemann, UK, 2007.

Beer, F.P., Johnston, Jr., E.R.: *Mechanics of Materials*, McGraw-Hill, Inc., New York, 1981.

Bela G. Liptak: *Instrument Engineers'Handbook,* CRC Press, London, 2003.

Burghardt, Henry D.: *Machine Tool Operation Part I: Lathe*, Forge. Bench, 1996.

Cairncross S, Feachem, R.G.: *Environmental Health Engineering in the Tropics: An Introductory Text*, Wiley, Chichester, 1993.

Callister, Jr., William D.: *Materials Science and Engineering – An Introduction*, John Wiley and Sons, UK, 2000.

Chen and Lui: *Handbook of Structural Engineering*, CRC Press, NY, 2005.

Chiasson, A.D.: *Advances in Modeling of Ground Source Heat Pump Systems*, Oklahoma State University, Oklahoma, 1999.

Cundall, P. A., and Hart, R. D.: *Numerical Modeling of Discontinuities,* Pergammon Press, ML, 1993.

Cyril F. Perry: *Relief Systems Handbook. Institution of Chemical Engineers,* London, 1992.

Dyer, Herbert J.: *How to Work Sheet Metal,* Model and Allied-Argus, UK, 1963.

Dym, J.B.: *Product Design with Plastics: A Practical Manual*, Industrial Press, New York, 1983.

Edgertpm, D. E. H.: *Industrial Research and Innovation in Business*, Elgar, Cheltenham, UK, 1996.

Emmerson, George S.: *Engineering Education: A Social History*, Crane, Russak, New York, 1973.

Fraenkel, P.: *Water-pumping Devices: A Handbook for Users and Choosers*, Intermediate Technology Publications, London, 1986.

Franceys R, Pickford J, Reed, R.: *A Guide to the Development of On-site Sanitation*, World Health Organization, Geneva, 1992.

Hammond, John Hays: *The Profession of Engineering,* John Wiley & Sons, Inc., New York, 1929.

Hans D. Baumann: *Control Valve Primer:A User's Guide,* Instrument Society of America, 1998.

Hicks, T.G. and Edwards, T.W.: *Pump Application Engineering,* McGraw Hill, NY, 1971.

Hodges, Henry: *Technology in the Ancient World,* Pelican-Penguin, London, 1970.

Jeffrey TD: *Hydraulic Ram Pumps: A Guide to Ram Pump Water Supply Systems,* Intermediate Technology Publications, London, 1992.

Kaplan, OB: *Septic Systems Handbook,* Chelsea, MI, Lewis, 1991.

McCreight, Tim: *Practical Casting: A Studio Reference,* Brynmorgen Press, Boylston, MA, 1986.

McNeil, Ian: *An Encyclopedia of the History of Technology,* Routledge, London, 1990.

Mujumdar, A. S.: *Advances in Drying,* Hemisphere Publishing, New York, 1980.

Otis RJ, Mara, D.D.: *The Design of Small-bore Sewer Systems,* World Bank, Washington DC, 1985.

Perry, R.H. and D.W. Green: *Perry's Chemical Engineers' Handbook,* McGraw-Hill. New York, 2007.

Practical Goldsmith: *Mounting - Joints, Electroplating,* Ruhle-Diebener-Verlag, Stuttgart, 1993.

Shigley, J.E.: *Mechanical Engineering Design,* McGraw-Hill, Inc., New York, 1972.

Stepanoff, A.J.: *Centrifugal and Axial Flow Pumps,* John Wiley and Sons, NY, 1948.

Timoshenko, S., MacCullough, G.H.: *Elements of Strength of Materials,* Van Nostrand Co., New York, 1957.

Turner, J.H.: *Small Engines: Operation, Maintenance and Repair,* TAB Books, Blue Ridge Summit, PA, 1987.

Whelan, Tony: *Polymer Technology Dictionary,* Springer, London, 1994.

Yam, K. L.: *Encyclopedia of Packaging Technology,* John Wiley & Sons, NY, 2009.

Young, C.B.F.: *Chemistry for Electroplaters,* Chemical Publishing, Brooklyn, 1945.

Young, R.A.: *The Rietveld Method,* Oxford University Press & International Union of Crystallography, Oxford, 1993.

Zappe, R.W.: *Valve Selection Handbook,* Gulf Professional Publishing, UK, 1998.

Index

H

I

J

K

L

M

N

O

P

R

S

T

W

❑❑❑